Readers' Reviews

Readers' reviews across generations and backgrounds brought great satisfaction. Some readers wrote to me how the book, through its effortless reading and humour helped them live through difficult times of tragedies and depressions.

Brig Beant Parmar, (Retd), Marathas, described this book as "an autobiography of every officer of the Indian Army. Every officer, serving or retired, must have a personal copy. You have done the Army a favour by writing it."

One veteran wrote, "My morning walk old soldier companions have loved it. One of them almost got thrown out of the house by his wife for laughing and sniggering all by himself." Many described it as "a collector's item."

A highly decorated war veteran had this to say:

"This book should be issued to all first termers in the NDA. The divisional officers must read select anecdotes from time to time to cadets to let them know how they are being trained to become leaders-physically fit, mentally alert and morally upright. Their stay at the NDA will be more enjoyable and meaningful... Your one tenure as the commandant in NDA will further add to the prestige and glory that NDA has attained as one of the best military training institutions in the world."

Brig Ranbir Singh (Retd), *AVSM, Vir Chakra, Defence Colony, Meerut*

And a few others are listed below:

"I had the opportunity of reading this famous book "Militarily Crazy..." I just could not keep it down, no matter what. Quite often my wife woke up with astonishment in the middle of the night, hearing me burst out in laughter. A beautiful and a healthy read."

Major Basant Sewada, *Armoured Corps*

" Written in an evocative style, brings up lots of memories of NDA onwards. One of the few books that describe military life-warts and all, in a very heart warming way. Would like to present it to a friend, I know he will have as much pleasure reading it as I did.

Col Harjeet Singh (Retd), *SikhLi. Author*

"The book was more than bringing a smile and cheer. I was always a little fascinated with the Army. After reading the book, I know I will give it the best shot. I too now have a dream to wear the 'Olive Greens' and write about my life, the way you did. Thank you. Hope you will reach even higher ranks, perhaps the COAS someday."

Vinit Vatsal *(Engineering student)*

I do hope, Honourable Raksha Mantri is listening and puts him on the selection board panel. Thank you for the compliments.

There were those who never wanted the book to finish, so read it slowly, in instalments. There were others who found it 'un- put-downable' and finished it over night. The list is too long to include all here.

The response has been heart warming, and it has raised my spirits higher than the "Old Monk" ever did. If it has brought cheer to even one depressed person, my effort has been worth it. It is this response that has prompted me for the second edition. I hope this time it will reach out to civil fraternity too and bring smiles and happiness to many more.

Maj Gen Anil Sengar

MILITARILY CRAZY

The Lighter Side Of Life In The Indian Army

MILITARILY CRAZY

The Lighter Side Of Life In The Indian Army

By

Maj Gen Anil Sengar, YSM, VSM

PENTAGON PRESS

Militarily Crazy: *The Lighter Side Of Life In The Indian Army*
By Maj Gen Anil Sengar, YSM, VSM

First Published in 2012

Reprinted by Pentagon Press 2014

ISBN 978-81-8274-783-8

Published by
PENTAGON PRESS
206, Peacock Lane, Shahpur Jat,
New Delhi-110049
Phones: 011-64706243, 26491568
Telefax: 011-26490600
email: rajan@pentagonpress.in
website: www.pentagonpress.in

Branch:
Prime Arcade
Office #11
1154 Saifee Street
Opp. M.G. Road, Camp
Pune-411001
Email: pentagonpresspune@gmail.com

Photographs
Courtesy: National Defence Academy
Indian Military Academy

Printed at Aegean Offset Printers, Greater Noida

Contents

Mani Shankar Aiyar
MEMBER OF PARLIAMENT
(RAJYA SABHA)

12, Safdarjung Lane
New Delhi- 110 011
Tel. 011-23795402
E-mail: msaiyar@hotmail.com

Foreword

"Militarily Crazy : The Lighter Side Of Life In The Indian Army" is a delightful reflection of the lighter side of soldiering spun together with great imagination. I have an emotional connection with the army myself. My father-in-law was an army officer who made the supreme sacrifice on River Irrawaddy in Burma in the Second World War. He and General Ayub Khan, the former President of Pakistan were buddies and fought the war together as young officers. Gen Ayub remained in touch with my in laws throughout his life. I therefore feel the vibes of pride and passion that the men in uniform demonstrate in their profession. As a career diplomat, I had the privilege of knowing service officers in different armies which has given me a good sense of their profession.

Maj Gen Sengar unravels the unknown facets of lives of soldiers through amazing real life stories written with passion. I remember a Pakistani Brigadier narrating me this incident. During the Indo–Pak War of 1965, his formation intercepted the communications of the Indian formation across the border. He recognized the commander on the Indian side. Both of them were buddies at the IMA, Dehradun, during training. He got on the radio, exchanged compliments and over a good laugh, said, "You so and so. You are opposite me in this war. I had beaten you in everything in the academy, I will beat you here too." It is typical of the armies and this book is an excellent reflection of that spirit.

The Indian Army man, constantly in combat, demands very little, is highly dedicated and above all, extremely loyal to his *'naam, namak aur nishaan'*. It is their pride in their profession that is the hallmark of the Indian Army. The book captures this passion and spirit of soldiering through this absorbing and humorous collection. I am certain readers from all walks of life will find it captivating. There are loads of wisdom in the letter of Field Marshal Cariappa and the experiences of Field Marshal Manekshaw for everyone to learn from. Soldiers of all vintage will find the book an enthralling journey down memory lane.

I compliment the author for putting together a fascinating collection that brings to life the vast canvas of military life. I agree with Maj Gen Sengar when he says, it will bring you cheer and smiles and start your day in good spirits. It certainly did that to me.

Jai Hind

(Mani Shankar Aiyar)

DEDICATION

This book is dedicated to the gallant Indian soldier. He asks for nothing. By his indomitable spirit and sacrifice, he provides a blanket of freedom and security under which the country sleeps peacefully, unaware of the hardships that he and his family endure.

To the infantryman, I tip my hat even-so-more. From amongst all those, who were privileged to don the 'olive greens', no one has done more or suffered more in service to the nation, in war or in peace than the humble infantryman. No matter what, he shall always remain the torch bearer of the Indian Army.

Dear Soldier, the country owes you a lot of gratitude for, ***"Never Have So Many Owed So Much To So Few"***

God Bless the Indian Army,
God Bless India

INDIAN ARMY IN SERVICE OF THE NATION

SIKKIM EARTHQUAKE: 2011

NO BRIDGE TOO FAR

Courtesy E-mail from Lt Gen RK Kadyan (Retd.)

PREFACE

Soldiering, "the profession of arms", is one of the oldest, if not the oldest of professions and the most honourable of all. There aren't many takers for this profession today because of the lure of mega bucks outside, not to mention the unwillingness to bear the rumble and tumble this profession entails. The whole concept of service before self, self sacrifice, camaraderie and brotherhood, honour-code and country and Chetwode motto, which guide the Armed Forces, makes it a truly special institution. Today, beyond the confines of the military cantonment these values are hard to come by.

The army is a unique organization and I have enjoyed every bit of it during my three decades of service. I have not been a "joker" in the military sense of the word, but I could pick the subtle, funnier side of military life and retain it somewhere in my brain's hard disk. The typical army way of doing things and the peculiar traits of its diverse troops make for a wonderful environment, where every day is worth its weight in gold. Over the years, whether in peace or operations, I have had many unforgettable moments. I was thus tempted to compile them into a book, so as to share my wonderful experiences with others. The book 'Militarily Crazy : The Lighter Side Of Life In The Indian Army' is not about anecdotes alone-it will also throw light on many unknown aspects of a soldier's life.

The incidents and events narrated in the book are generally speaking true as I have been witness to most. Some others have been narrated to me, based on others' experiences. There may also be anecdotes, whose veracity cannot be established, but have been around in the army folklore for long.

All names and characters in the book are changed,

resemblances if any are coincidental. Where the characters are indicated as actual, it is with their approval.

Often, I have narrated these experiences and they have brought cheer and smiles to many. I have no doubt many more will find a reason to start or end their day with a smile. I guarantee it will.

As I see it, one basic principle of humour is that we laugh with others and not at others. Anecdotes on a Regiment or class of people are based on prevalent stories. Nothing in this book is meant to denigrate an individual, Unit, Regiment or a class of people.

The book contains episodes from my experiences as a cadet in National Defence Academy (NDA), Indian Military Academy (IMA) and reminiscences from various facets of military service that go into making our Army wonderful. The effort started in 1992 and took nearly twenty years to complete.

The NDA that I talk about in the book was one of 60s and 70s. Things have changed since; for good or for worse, I am not so sure. The NDA was known for its ragging and typical flowery language invented, nurtured and evolved painstakingly over half a century. This was a cultural shock to the new arrivals, but they got used to it soon as they had no choice. The cadets emerged from the Academy as a physically and mentally tough lot. They owed this, not so much to the Academy training, but to the rigours that they were put through during the rest periods, in sweltering heat and rain. Ask an ex NDA and you will know how proud he is to have gone through that mill.

It took the cadets sometime to get used to this and they realized that all those mind boggling abuses, which were rapid fired like a machine gun, meant no harm. The language gave the academy its character and it would not be the same without it. The feel of the prevailing environment during a cadet's life would be lost without its "lingua franca". Nevertheless, that part has been completely toned down in the book or else it would not have seen the light of the day. That is a compromise, I wish I did not have to make.

As you leaf through the book, you will get a good sense of what these wonderful academies are all about and what it takes to pass out from them in one piece. It will also give you a feel of the regimental life of an officer as he grows from a Lieutenant (Mister) to a Battalion Commander (Gaon Burha). It will give you an insight into the whole concept of soldiering and what makes our soldiers and officers, the winners that they are.

I have no doubts that this book will help you start and end your day with a smile. Remember what Charlie Chaplin said, "My those days are wasted, when I have not laughed."

I need to thank a whole lot of people who have contributed in different ways in compiling the book-Shiva with sketches, Additional Directorate of Public Information, Army, for their assistance during planning of the book, and in particular, Colonel Bhasker Gupta for his extremely useful suggestions on a variety of aspects. I would like to express my grateful thanks to Shri Mani Shankar Aiyar, Honourable Member of Parliament (Rajya Sabha), for the foreword. We often hear him engage in serious and intellectually stimulating debates. A brilliant orator, he also has the finest sense of humour that I have witnessed. This, and his emotional connection with the army prompted me to approach him for the foreword which he most readily obliged with. I owe my special thanks to my battalion, 5 GUARDS. It is the experiences with those wonderful officers and men in tough times where the idea of this book originated. I also need to thank **'this wonderful army of ours'** which has helped me to live a life of pride and dignity.

KNOW YOUR ARMY

Services are the holy cow, little known, even lesser understood. The country needs to know and understand its army, not fear or envy it. As a young officer I was asked by a young lady, "How do you pass time in peace when there is no war?" In 2001, a seminar on 'Jammu and Kashmir : The Way Ahead' was conducted in Jammu University, attended by eminent speakers. A gentleman, if I could call him that, during a discussion on threat to soldiers' lives in insurgency commented to his colleague, "So what? The army is paid to die."

These instances demonstrate ignorance of many who have no idea what soldiering entails.

Why army deserves a privileged place in the comity of institutions should not be an issue of debate at all. Nevertheless, the following unknown facts will help the environment understand the army better.

- It is only in the military that an officer is commissioned by the President under his signature. It reads "I, ***reposing special Trust and Confidence in your Fidelity, Courage and Good Conduct...***" An honour no one else has, justifying the sobriquet, 'the noblest profession'.
- No other service is based on the ethos of giving your life for others, your organization and for your country. In no other service does an individual or his family suffer so much for so long.
- Armed Forces are your last bastion; if it collapses, the country will collapse.
- The country can survive any inefficient institution, it simply cannot afford a disgruntled armed forces. Ten disgruntled

soldiers could become a cause of national embarrassment.

- From the rescue of Master Prince from a bore well, to national security, the country looks at the army as the messiah. It is the army that pulls out everyone's chestnuts from all kinds of fire. And they never even claim credit for what they do.
- Never have so many owed so much to so few.
- No institution is respected half as much for its ethos and values or its contributions as the military. I was standing at the Amar Jawan Jyoti at India Gate in uniform with a few officers. Hordes of people, children, young couples and old folks walked up and said, "Sir, can we have the honour of a photograph with you." We were no celebrity; they wanted to be photographed with Indian Army Officers. One young mother stood beside me for a photograph with a baby and said, "Can you hold him in your hands? He will be blessed and God willing, he will be an army officer one day."

The army is all about teamwork, loyalty, honour, self-respect, self-sacrifice and camaraderie. You trust your mate with your life. We would risk four soldiers to save one soldier's life. We would never leave the body of a fallen soldier behind, come what may. A soldier will die but will not allow his weapon to be separated from him, for that is a matter of honour. He does duty for the honour of his unit, the famous ***"Naam, Namak aur Nishaan".*** His unit is his home, and his colleagues - his band of brothers. He prefers death to defeat. To us, the men in uniform, the reverse of 1962 war hurts us even today, even though we were not even born then. Does it to others?

Unmindful of the degenerating moral and ethical standards all around, army proudly marches on its value system, conspicuous by absence elsewhere. A soldier is motivated by the honour and the colour of his lanyard. Money figures nowhere in his calculation.

A few incidents describe the culture, ethos and values of

our army that must be mentioned. During the Indo-Pak War of 1971 in East Pakistan, one battalion captured a Paki position with a whole lot of prisoners. It was bitterly cold. The next day, an Indian JCO asked the Pakistani JCO to provide blankets for our officers and men from their stores. The Paki JCO was shocked to learn that Indian officers were out in the open the whole night with their men in bitter cold with nothing. He was surprised that –that was the way Indian officers operated and that there was nothing exceptional about it. He is known to have commented that, if Pakistan had Indian officers, the story of the war would have been very different.

I will narrate my personal experience on two occasions with Paki troops. On a UN mission in Cambodia in 1992-93, my clerk was a Paki NCO. I would see him commute daily to and from the office on a rickshaw. The Pakistani contingent at Phnom Penh, officers and men, were staying in one hotel. One day I asked him, why he commutes by rickshaw when his officers come to the same office from the same hotel in a UN car. His reply, they are not like Indian officers. They do not take us in their vehicle, even though we come from the same hotel to the same office. He added, he sees the Indian officers bring their NCOs in their vehicle every day, sitting alongside in the co-driver's seat.

On another occasion, I was posted as Defence Attaché in a foreign mission. A Pakistani JCO had been posted in their embassy for over two years. He had interacted with me on many occasions and told me the same thing in front of his countrymen. If the Pakistan Army had Indian officers, they would be an unbeatable army. That is the best compliment that one can get on leadership. In the Kargil War of 1999, the ratio of fatal casualties of our officers to men was about 1:8, whereas the ratio of officers to troops in a unit is over 1: 60.

With the proliferating media, a soldier is fully aware of the environment in the society. Notwithstanding the army ethos and discipline, coming from the same society with its ills, he can not be left unaffected for long. A soldier of today is politically and socially aware and a discerning individual.

I served a tenure in a foreign mission concurrently accredited to two more countries. The experience only reaffirmed my belief in the yeoman service the army does for a pittance, based on values which are alien to others.

While this book is all about the lighter side of the profession of soldiering, it is a bloody serious business. Nowhere else are the consequences of incompetence and failure as severe as here, both in terms of human lives and for the nation. A lost war becomes an indelible part of your history, forever. Services are our last instrument of national policy. They need to be looked after, kept sharp and given their due. The country will do well to remember that peanuts attract only monkeys and monkeys won't win you wars.

I visited the famous Gallipoli peninsula in Turkey. Almost twelve hundred Indian soldiers had made the supreme sacrifice there during the First World War. The Turkish Army had stopped the combined might of Great Britain with troops from India, Canada, ANZAC and France. Turkey lost over seventy thousand troops in the process. In memory of their soldiers the nation has made a massive memorial, largest that I have seen, visible from tens of miles. Standing beside that towering memorial, I stood with an elderly Turkish person. I commented, "This is a real big memorial." The old Turk replied "It is big alright, but not big enough for the sacrifices of our soldiers." That is the spirit of a nation that cares for its soldiers, for every able- bodied Turk serves in the army for some time. We would be a better nation if we could do that too.

The Turkish President, the Supreme Commander of Turkish Armed Forces, reviews officers passing out parades of the three service academies every year without fail, two in one day alone. Nothing else can happen on that day. Similar gestures in India will be good for our nation as well.

Some wise man has said:
'Maintaining an army is an expensive business. If money is spent on the army and the army is not used, you only lose money. If you do not have a competent and a well equipped army, and it does not deliver when needed, you lose both, the money and the army.'

I would like to add to that- *the national pride.* And as I said, it remains an indelible part of your history- forever. And it hurts.

For India, a good army is not an option, it is a compulsion and I do not see that changing anytime soon. While machines are important, it is the men behind the machines that win wars; Kargil demonstrated that clearly. They better be motivated and good and for that the government must do all that is necessary.

Nothing conveys this sentiment better than the advice to his king by India's most celebrated thinker from history 'Chanakya':

The great Prime Minister of Magadha, Chanakya, told Emperor Chandragupta Maurya :

"Pataliputra rests each night in peaceful comfort, O King, secure in the belief that the distant borders of Magadha are inviolate and the interiors are safe and secure, thanks only to the Mauryan Army standing vigil with naked swords and eyes peeled for action, day and night, in weather fair and foul, all eight praharas (i.e. round the clock), quite unmindful of personal discomfort and hardship, all through the year, year after year. To this man, O Rajadhiraja, you owe a debt: please, therefore, see to it, suo moto, that the soldier continuously gets his dues in every form and respect, be they his needs or his wants, for he is not likely to ask for them himself. The day the soldier has to demand his dues will be a sad day for Magadha; for then, on that day, you will have lost all moral sanction to be the king! "

It is for the well being of the country that someone else must think about the soldier.

PART I

PEARLS OF WISDOM

Indian Army has had two Field Marshals. Late Field Marshal KM Cariappa, OBE, then General, was the first Chief of the Indian Army (15 Jun 1949 -14 Jan 1953) on attaining independence. He steered the Army during its infancy when it was tragically split on partition. Comrades in arms, who had fought side by side, suddenly became adversaries. General Cariappa laid the foundation of a proud, professional and a dedicated army. The nation bestowed on him the honour of 'Field Marshal' in 1986 in recognition of his yeoman service to the Army and the Nation.

Here, I have chosen to include excerpts from Field Marshal Cariappa's letter written to every officer. Written over sixty years ago, it is as relevant today.It indicates that the rules of conduct of an officer do not change. They are based on the finest principles of behaviour as an officer and a gentleman; all in the best interest of the country and its people.

Late Field Marshal SHFJ Manekshaw, MC, led India to its most glorious victory against Pakistan in 1971. Memorable recollections of Sam Manekshaw from his talks at different times and places have been narrated here. He was known as Sam Bahadur, being a staunch Gurkha officer. In the Army he was popularly known as Sam, and was gifted with an excellent sense of humour.

While the first, the serious one, is meant to reiterate the principles of conduct as laid by our first Chief; the second, will reflect on various facets of personality of another Chief who led the Indian Army through another challenge of a different kind.

FIELD MARSHAL KM CARIAPPA, OBE

(Excerpts from the first letter written by Field Marshal Cariappa to all officers)

I send all officers of the Army the following message. I earnestly hope that every officer in the Army today will take all that follows in the spirit in which it is written, and live up to the requirements stated, hundred percent in the service of our country.

Welfare of Troops

Our troops have always been magnificent in their sense of loyalty to the Army and the country. We can ensure that this satisfactory state of affairs continues if we keep our troops happy and contented in every way. This calls for personal attention of every officer, to the welfare of the men placed under his command. Their needs are simple. If the jawan feels that he has an officer who is professionally efficient, a gentleman and a person in whom he can place his confidence to safeguard his personal and domestic interests, we officers will have gone a very long way to ensure that man's loyalty to the Army and our government.

As I have often said, our jawans are absolute gems. (In the term 'jawan' I also include our clerks, our chaprasis and our sweepers). I would like you all to make everyone of them feel that they are proud to serve under you as a Commander, whatever your command may be, however big or small it may be.

Civilians

We must make them all feel that we, as officers, are always ready to associate ourselves with them as citizens of the land in our common efforts to serve our country, in our respective spheres of activities. Please make quite certain that you do NOT lay yourselves open to criticism by our people for lack of courtesy and good manners in your general conduct both in public and in private. Pay due respect to civilians senior to you and be as courteous as you possibly can, as an officer and a gentleman, to those above you, equal to you and below you in civil life, for courtesy and good manners do NOT cost you anything. Let us set an example to the rest of our country in regard to the value of these two essential things, i.e. courtesy and good manners.

Politics

Please do NOT get mixed up with party politics. Officers and soldiers should NOT take an active part in political matters. You will appreciate that if we did so, we will not be serving the government of the day loyally. Remember we serve the people of the country and, therefore, the government of the day is that government which is put in power by the people. This, however, does NOT mean that we, as citizens of the country, should neglect our understanding of various political parties and their objectives. Now that we have a right to vote, it is imperative that all of us, officers and men, should be quite clear in our minds in regard to exercising our privilege of voting.

Economy

So, every one of us in the Army must do everything we can to save every pie we can, so as to ease the present financial burden on our government. Economy of transport, petrol, supplies, equipment, accommodation, personnel will all go to help solve this problem. See that not one of these is used more than is absolutely necessary and essential to ensure the efficiency of the Army. Remember, as I have often told you before, we are the custodians of the tax payer's property. Everything in the Army such as the service of the personnel, supplies, equipment, accommodation, vehicles, animals etc all belong to him. It is our sacred duty to see that we do not misuse any of these for our personal use in any way whatever. The proper care and storage and the efficient maintenance of all store, weapons, equipment and vehicles is our personal responsibility.

Public Funds

Every officer must make quite certain that Public and Regimental funds are not misappropriated or embezzled. I would ask you all who are placed in charge of public funds, including your Regimental Funds of all kinds, to be scrupulously honest in the handling of these funds. Do not be tempted to use these funds for private purposes.

Ex-Servicemen

Firstly, please deal sympathetically with all cases of ex-servicemen which come to your knowledge. If an ex-serviceman comes to you with a request for help, do not send him away by saying, "No, this is NOT my job." Tell him, if it is not your job and you cannot help him, where he can go and whom he should see for help. Remember, these ex-servicemen have done their stuff for the country. It is because of their grand work in the past, coupled with our humble efforts since we have been free, that our army today enjoys the respect, the confidence and the affection of our people. We cannot neglect the interests and welfare of our ex-servicemen. Go out of your way whenever possible to help them if you can.

Professional Study

Your profession today is a very complex matter. The advent of mechanization and scientific warfare has increased your responsibility in regard to your being prepared as an efficient soldier and officer in the service of your country. It must be your constant endeavour to see that you are up to date in military teachings in so far as they affect, initially your own arms of the service, secondly the Army and thirdly the three services, Army, Navy and Air Force as a whole. This calls for very extensive and continuous study. Do NOT waste your valuable time on useless pursuits. Plan your time in such a manner that you allot adequate portion of it for study without forgetting time for recreation and relaxation. You must take part in the organized games, both individually and with your men.

If we are to keep pace with modern developments, we must give the question of scientific research and operational research our very serious thought.

You must concentrate on ensuring that the basic training of the men you command is thorough in every respect. Do NOT be slip-shod in this respect. This means your spending more time on study and training. Do NOT waste time on useless pursuits, for every second you so waste in peace, you will pay for in blood in times of war, for the lives of the men you command are in your hands and lives are too sacred to be so used. In your training, pay very great attention to minor tactics because if your training in minor tactics is really sound and thorough, your problems in action will be made so much easier.

Officers' Private Life

I would like you also to make quite certain that you do not over-do your amusements. By this I mean, do not indulge too much in lavish entertainment, too much dancing and things like that. I am very sorry to see that the object of club life is NOT being properly understood by a number of officers in the Army today. Please do NOT be under the impression that Club Life is designed mainly for DRINKING and DANCING. It is not incumbent upon you to drink alcohol when you go to the club. Remember, alcohol is expensive these days. Think of your children's education or of putting by some money for a rainy day.

Officers' Mess

I would like to see the Mess life kept up with all its dignity, propriety and correctness. Do NOT make your messes cheap by inviting all and sundry to your parties. I regard Messes as excellent training grounds for officers of all ranks from the point of view of maintaining discipline, comradeship, teamwork and building up one's character. Make quite certain the Mess is treated as a home for the officers of the unit.

Officers must be dressed properly at all times. Please also ensure that not only your ante-rooms are kept neat and smart but your pantries, kitchens, and washing places are also kept equally smart and tidy.

As regards food in the Mess, it must be simple but wholesome, well cooked and properly served. I do not think it is attractive to see the table covered with bottles of all kinds of sauces, sliced onions, chutneys and pickles dotted about. All these can be kept on a side table and brought up as and when required.

I wish to see Messes used, as I have said above, as a home for officers. Whilst maintaining discipline and propriety in regard to paying respect to senior officers and visitors, I do not wish to see Messes as parade grounds where officers click their heels and stand to attention whenever they talk to a senior officer.

Security

You have often heard the old saying, "Careless talks cost lives." Every one of us has a very sacred duty to the people of our country whose security we have pledged to guard. Please be hundred per cent security-minded at all times. Do NOT talk loosely in public or even in private to show off, "I am in the know," merely to gain cheap popularity.

I have already put over to you all, through your Army Commander, the desirability of not consuming alcoholic drinks at public parties. You know the reason why. You see some of us can take drinks and some cannot. It is amazing how one's tongue gets loose after a glass or two of whisky or some such other strong drink. I have heard, much to my horror, some getting a bit too talkative on service matters at public parties after they have had a few drinks. Surely, we can do without these drinks at these public parties when refusing drinks is indeed in the interest of the security of our people. It is not a matter to be ashamed of to tell your hosts when you go to a public party that you do not wish to take any alcoholic drinks. On the other hand, if your hosts know of the noble idea behind your refusal to take a drink you will be held in a greater esteem by them.

Commissioned Officers

We are all members of one and the same team where good comradeship, fellowship, loyalty to each other and cooperation between all ranks at all levels, are absolutely essential for the good of our Army. Please do NOT get into the despicable habit of becoming ranks conscious. Take what job is given to you. Do NOT ask for jobs. Give your best, your very best, at all times. We have all got to do this if we want to serve our Army and our country well. I will ask you all again, please, NOT to indulge in another despicable habit, which is criticizing senior officers and indeed criticizing junior officers also. If we all did our normal work conscientiously and completely, no one can have a single spare moment left to indulge in such talk about each other.

Junior Commissioned Officers

I would like every officer to see that Junior Commissioned Officers, who are the backbone of our Army, do receive their proper recognition as such and are given every facility and guidance to carry out their job well. They must continue to receive from every officer the same degree of courtesy and regard and respect as they used to enjoy before the War. Some people seem to think because we are a fully nationalized Army, there is no place for JCOs in the Army. They argue that JCOs were a necessary link between the rank and file and the officers during the British regime and that such a link is not essential now as there are no more British officers in our Army. People, who talk thus, miss the most important aspect of this matter. The JCO has always been, and always will be primarily a commander of a sub-unit under the command of a commissioned officer. These appointments, i.e. commands of sub units have been held by our JCOs most efficiently throughout the history of our Army. They withstood the test of two of the biggest wars fought in the last 200 years. I would like every officer to regard these appointments in the manner I have stated and in no other way. Such JCOs who are considered fit to be promoted commissioned officers, have been in the past and will in the future, be so promoted. They will be given every help to ensure that they enjoy their full privileges as commanders of sub-units as they did in the past. Every officer in the Army will see that the izzat of our JCOs is maintained at all times.

Finally, as long as you remain in the Army, whether you are a regular or temporary officer, please give your very best in the service of the country until the very last second you are in the Army, working as a loyal, well disciplined and honest member of this big team, our Army, as an Indian and with no anti-communal feeling of any kind whatever. Please do not allow the virus of communalism to get into the Army. Communalism is a most destructive parasite, which if allowed to grow in the Army, will spell disaster for the country as a whole. We must maintain and enhance the good name we are fortunate enough to have today, which has been so liberally and generously expressed by our people as being a high factor which has helped and will help to provide the security and safety for our country.

I feel absolutely certain, with NOT an iota of doubt in my mind, that all officers in my team, of which I have the proud privilege and the honour today of being the Captain, will help me in the matters I have stated above to see that we have a first class Army in all respects.

Do nothing in private, that you will be ashamed to admit in public.
Field Marshal KM Cariappa,OBE

FIELD MARSHAL SHFJ MANEKSHAW, MC: THE LEGEND AND A SOLDIER'S SOLDIER

Field Marshal Sam Manekshaw passed away on 26 June 2008 at 94 after a long illness. Sam, as he was popularly known, had led the military to its most famous victory in the 1971 Indo- Pak war that led to the birth of Bangladesh. He was known as a legend and a great soldier, an inspiring leader and a great son of India, a truly great strategist. He will always remain an inspiration for our armed forces and the rest of the nation.

The towering figure that he was, I think, a brief narration of his experiences would be the right way to record for posterity.

Sam was a wonderful soldier gifted with an excellent sense of humour. He was a charismatic commander who led India to a lightning victory in Indo- Pak war of 1971. The credit for the glorious victory goes to all ranks of the Indian Military, who served there, led by Field Marshal Sam Manekshaw.

Here are some recollections from his talk to us during his visit to Defence Services Staff College, Wellington in 1990-91.

Quote, "Life can play games that one never can imagine. Before the partition I was posted as the Grade (1) staff officer in the Army Headquarter. My Grade (2) was Major Yahya Khan sharing the same room. When the partition took place, he chose to go to Pakistan. We had a casual chat about it and I had nothing to say on the issue.

When it was time to leave, he said to me, "Sir, I know that you have got a good motorbike. I am going to Pakistan, it is unlikely that I will be able to get anything of the kind for a long time. Can I take your bike? I will pay for it." I told him that I loved the bike a lot but I could not refuse him. He took it along to Pakistan.

He was to pay me a thousand rupees for it. He did not have the money then and promised to send it at the earliest.

Time flies. I had forgotten all about the bike and the money. Almost a quarter of a century later, I became the Chief of Indian Army and my friend Major Yahya Khan had done even better. He had risen to become General Yahya Khan, the military dictator and President of Pakistan.

A delegation was visiting Pakistan. I decided to get in touch with my old friend Yahya Khan and wrote him this letter stating :

Dear General Yahya Khan,

So much time has passed since we parted company in 1947. I have become the Chief of the Army here in India and you have graduated from Major to President of Pakistan. I thought I must remind you Mr. President that at partition you had taken my motorbike for a thousand bucks. Twenty five years later, you still owe me that amount. I believe, now that you are the President, your financial situation will enable you to clear the long standing due. I wait eagerly as my resettlement plans hinge precariously on this money.

Regards,

President Yahya Khan replied:

My dear General,

Thank you indeed for your letter. It is indeed a great pleasure to hear from you after an eternity, reminds me of the bygone days. With regard to your thousand rupees for the motor cycle, I must mention that I very much remember the debt. I, however confess, that my financial state today is no better than when I took the motorcycle and find myself unable to pay the money at the moment. I promise to return the money at the earliest.

Regards," Unquote.

The promise remained unfulfilled.

Tragically, it is a twist of destiny that the two who worked in the same office, led their armies on the opposite sides of the fence in the 1971 war. General Yahya Khan died later. Sam harboured no hopes for his thousand bucks, but often narrated this wonderful story.

Sam and the 1971 Indo- Pak War

The Pakistani inability to accept the election victory of Sheikh Mujibur Rahman of East Pakistan, finally led to the genocide in 1971 setting the stage for creation of Bangladesh. It had created a serious situation forcing millions of refugees to pour into India. For India itself, the situation became extremely critical. When the world was paying no heed to India at the time of crisis, the belligerent attitude of Pakistan had forced India to act in its own interest. It was decided to restore the situation so that millions of refugees could go back. Bangladesh had already launched its war of independence.

In various meetings chaired by the then Prime Minister Mrs. Indira Gandhi, General Sam Manekshaw was asked to launch operations within a month. Sam knew the operational preparedness of the army better and said that operations could not be launched earlier than nine months. That was not acceptable to Mrs. Gandhi, who insisted that the operations be launched no later than three months. Sam would simply not agree and stated that the army lacked the wherewithal to go to war and would need at least nine months to prepare.

The Defence Minister then was the inimitable Babu Jagjivan Ram. He would take Sam to a corner and try and convince him without success.

Another high level meeting after some days and the same

effort by the Prime Minister to convince the Army chief but Sam Manekshaw stood his ground. Defence Minister Babu Jagjivan Ram then told General Manekshaw, *"Arre Shyam mann jao na."* (Oh Shyam, please agree).

General Manekshaw in his typical style said, "First and foremost, Babu ji, I am Sam and not Shyam. And for the war, it was not a matter to bargain, we were either ready or not ready."

Mrs. Gandhi then asked Sam, "How much time would India need to liberate Bangladesh if we went to war after nine months?" Sam replied, One month. The war was over in two weeks. Later Mrs. Gandhi asked him, "Sam, before the war when I had asked how long will you take, you had said one month, but you took only two weeks?"

Sam replied "Even then I knew that we would wind up in two weeks. But I did not say so, because if I had said so, you would have been breathing down my neck after seven days."

On another occasion after the war, when there was a wild rumour of a possible military coup, Mrs. Gandhi asked him if he was preparing to take over the country. Pointing to his long nose, he said, "I know I have a long nose but I do not use it to poke in to other's affairs."

When Mrs. Gandhi asked him to go to Dacca to take the surrender of the Pakistani army, he declined magnanimously, saying that this honour should go to his Army Commander (Lieutenant General Jagjit Singh Arora) who was the architect of this victory. He would only go if it were to accept the surrender of the entire Pakistan Army.

In preparation for the war, one of the problems facing the Indian Army was the shortage of ammunition for the Russian guns purchased earlier. Russian factories producing the ammunition for those guns had closed down. Sam along with a delegation from the Ministry of Defence went to Russia to explore the possibilities of procuring the much needed ammunition. Sam met his Russian counterpart. After a very cordial meeting, the Russian Marshal, saying "I know you would like to go shopping," took out a thick wad of roubles and offered it to Sam. Sam thanked him and politely declined saying he simply could not accept the money. The Russian Chief, a bit surprised, said, "But the civilian with you has already taken the money." Sam never mentioned the name, but with the war on the horizon, a fair guess cannot be too difficult.

Well, Sam went for shopping with his own money. In a shop, he was contacted by a stranger who said, "Aren't you the Indian Chief?" Sam said yes he was, and enquired about him.

He said, "I will tell you later, but I understand you are here in search of artillery ammunition for your guns."

Sam said yes indeed he was, but who was he. He said he was an Israeli businessman and that he would provide the same ammunition at half the cost. A meeting was fixed with him. The issue was discussed with the ministry representative who declined to be a party to this arrangement as India did not have diplomatic relations with Israel then. Sam said, "You no tell, me no tell." India fought the 71 War with Israeli ammunition procured at half the cost, courtesy Sam.

One of Sam's sayings which has gone into Gurkha Regiment legend is "If anyone tells you he is never afraid, he is either a liar or a Gurkha."

Sam was a forthright commander who stood up to Mrs. Gandhi and confronted her on the 1971 War. Lieutenant General Depinder Singh, his MA from 1969-73 described him as "A banishing league of Generals, courageous enough to defy political bosses". The PM and the cabinet pressed him for an offensive in East Pakistan in Apr 1971, but he held his own. He said, "He was not prepared for a war in Apr as the passes on the Himalayas were opening and a Chinese aggression could not be ruled out."

Sam used to say : "There is a very thin line between being dismissed and becoming a Field Marshal." Sam had his way as the war took place seven months later giving time to the troops to prepare. He told the prime minister after the cabinet meeting, *"It was my job to tell you the truth. And it is my job to fight and win."*

The war commenced in December and the army wrote a glorious chapter with its outstanding achievement, liberating Bangladesh in a fortnight. The war over, more than 95,000 Pakistan troops had surrendered to the Indian Army. The prisoner of war camps were established in different parts of the country. The families of the Pakistani soldiers were kept separately and the men folk were permitted to visit the families occasionally. A Sikh battalion was tasked to provide guards and security to the Pakistan family garrison. Soon Sam Manekshaw received a complaint that these boys had glad eyes and were playing the fool there. Sam said, "Change the battalion and put Gurkhas there. Not that they are any better, in fact in this regard they are worse, but they know how to go about it and with their 'Behni-Behni' approach they will never get caught and there will be no reports."

Sam passed away leaving a vacuum. The country will miss this wonderful "Officer and a Gentleman".

So that was our inimitable Sam. He shall forever continue to be the icon of the Indian Army.

NATIONAL DEFENCE ACADEMY
THE CRADLE OF LEADERSHIP

NDA PRAYER

O God, help us to keep ourselves physically strong, mentally awake and morally straight, that in doing our duty to Thee and to our country, we may keep the honour of the Services untarnished.

Strengthen us to guard our country from external aggression and internal disorders.

Awaken our admiration for honest dealing and clean thinking, and guide us to choose the harder right, instead of the easier wrong.

Kindle our hearts with fellowship for our comrades at arms and with loyalty to the men we command.

Endow us with the courage which is born of the love of what is noble and which knows no compromise or retreat when truth and right are in peril.

Grant us new opportunities of service to Thee, to our country and to the men we lead, and ever help us to place such service above self.

THE MAJESTIC SUDAN BLOCK

HUT OF REMEMBRANCE

LEST WE FORGET
THEY GAVE THEIR TODAY FOR YOUR TOMORROW

Academy Honour Code

"I Believe That A Cadet Must Be Loyal, Truthful,
Trustworthy, Honest And Forthright Under All Circumstances.
I Will Not Lie, Cheat Or Steal,
Nor Will I Mislead Or Deceive Anyone.
I Undertake To Faithfully Live Up To This Code And To
Continuously Encourage My Comrades To Do So."

Cadets' Commandments

The Nation	:	The Nation's honour and interest comes first, always and every time.
The Services	:	We pledge to maintain the traditions of the three services and keep the 'Trishul' always bright and strong.
The Academy	:	We are proud of the Academy. We will at all times endeavour to uphold its reputation by our conduct, bearing and manner both within and outside the campus.
The Corps of Cadets	:	We are a family. We are loyal to each other. We are as one man.
Truth	:	We will always follow the harder right rather than the easier wrong.

Discipline	:	We will obey all orders at once. We will never leave a task unless permitted to do so by a superior.
Punctuality	:	We will make punctuality our second nature.
Supervision	:	We will work without any need of supervision. Supervision is unnecessary when duty is performed conscientiously.
Work and Play	:	We will work hard and play hard. We will keep ourselves physically fit, mentally alert and morally upright.
Gentlemen	:	As future officers of the Armed Services, we will learn to conduct ourselves as gentlemen.

PART II

THE NATIONAL DEFENCE ACADEMY : THE CRADLE OF LEADERSHIP

The National Defence Academy (NDA) is the finest military academy in the world. Its architecture and its ambience is simply awe inspiring. Its age old traditions, some apparently silly on the face value, are the ones that all ex NDAs are proud of. It has given the country outstanding military leadership that has provided the country rock solid security. They believe that the character traits developed in the academy because of its ethos, tough regimen, camaraderie and esprit de corps makes them physically, morally and mentally robust. A 'never complain' attitude and the indomitable spirit makes them the perfect material to fight and lead men in battle. We believe that we were transformed from school boys to battle ready young men by a way of life at the academy that civilians may not comprehend. No wonder that NDA course 'get togethers' are common place even when they have one foot in the grave. In short, if you have not been through the NDA, you haven't been anywhere.

The ingenious tricks of the NDA were left behind as the cadets went to service academies. Those days, it was only when the cadets landed at the Indian Military Academy, Dehradun, did they realize what they had left behind. A cadet's day started with a prayer to uphold the honour and prestige of the armed forces and defend the honour and dignity of our country. Eventually, he will stand unflinchingly 'for the ashes of his father and the temple of his GOD' and many would have the honour to have their names inscribed in the 'Hut of Remembrance' and become immortal. It would be worth the while to see the evolution of this wonderful academy, the makers of men.

The Genesis

Commissioned in 1932, the Indian Military Academy (IMA), located at Dehradun, was the only academy training officers for the British Indian Army. Lessons from the Second World War brought out the inter-dependence between the three services and dictated the need for a joint services academy to train future leaders for combined operations. The vision of Lord Mountbatten in consonance with the sustained impetus and groundwork provided by Field Marshal Sir Claude J Auchinlek, C-in-C in India, laid the conceptual foundation for a Joint Services Military Academy modelled on the lines of the US West Point. A committee headed by Field Marshal Auchinlek, after extensive studies of various military academies around the world, submitted its recommendations to the Government in December 1946. The Government of India, therefore, approved the creation of an Inter-Services Wing for training cadets for the Army, the Navy and the Air Force together. Thus, in January 1949, the IMA was redesignated the Armed Forces Academy, with the military wing remaining on the present campus at Prem Nagar and the Inter-Services Wing established at Clement Town.

Birth of National Defence Academy

After independence in August 1947, this report was referred to the Chiefs of Staff Committee. Their suggestion for the formation of an interim Inter Services Wing at the Indian Military Academy, Dehradun was then implemented. A simultaneous action plan to commission a permanent war academy at Khadakwasla, Pune was also commenced and the foundation stone was laid by the first Prime Minister of India, Pt Jawaharlal Nehru on 06 October, 1949.

On 1 January 1949, the Armed Forces Academy having its Military wing, now called the Indian Military Academy and the Joint Services Wing (JSW) were commissioned. After two years of training at the JSW, army cadets went on to the Military wing for a further two-year pre-commission training. The naval and the air force cadets were sent to Dartmouth and Cranwell in UK for advanced training. On 07 December 1954, the interim process crystallized with the commissioning of the NDA. In December 1954 the Joint Services Wing moved from Clement town, Dehradun to Khadakwasla, Pune in Maharashtra, along with its name, formation sign and the Commandant. The NDA was formally inaugurated on 16 January 1955.

Why Khadakwasla? The NDA is located south-west of Pune City and north-west of Khadakwasla Lake on 7015 acres of land, out of the 8022 acres donated by the Government of the erstwhile Bombay State.

Pune was ultimately chosen after careful deliberations for its salubrious climatic conditions, suitability of terrain for military training, proximity to the Arabian Sea, existence of an operational airfield at Lohegaon, vicinity of military establishments and the presence of a lake nearby. The existence of a combined training centre and the 'mock' landing ship, HMS Angostura on the north bank of the Khadakwasla lake lent additional leverage to the claims of Khadakwasla over other contenders for the prestigious NDA. It is befitting that NDA is located in the happy hunting ground of the legendary Shivaji, with the daunting Sinhagad Fort as a panoramic backdrop.

(History of NDA. http://nda.nic.in/html/nda-history.html)

The NDA is the only tri service training academy of its kind in India, and may be in the world, I guess. It trains cadets of the Army, the Navy and the Air Force together for three years, who then proceed to respective service training establishments for specialized training. The benefit of a tri service organization is best reflected by the fact, that quite often, all the service Chiefs are course mates and half the issues get resolved without a fuss. In the tri service organizations that exist in the army, it serves well to have been in the same academy, either as contemporaries or otherwise.

A Walk Down Memory Lane

Known as the 'Cradle of Leadership', it is located on a sprawling unending estate in Khadakwasla, Pune. It has great traditions and customs. Mostly good, some silly and some even sillier, and all those have evolved and survived decades that make the NDA an unforgettable experience. If you want to join the Army, there is no better way than through the NDA. The spirit of NDA is all pervading. When two ex NDA cadets meet, the first question is, *"Which course, which squadron?"* And that wonderful bond is established. The bondage from the difficult times spent together builds up great camaraderie amongst course mates and squadron mates that lasts as long as you do.

With changing times, the NDA has changed too. In those days the academy was a difficult place to survive, where every day spent in one piece was an achievement. It fully subscribed to Charles Darwin's theory of 'survival of the fittest'. If you were clever enough to make up your losses or deficiencies by managing, known in the academy as the principle of *"Beg, Borrow or Steal"* (BBS), you were termed smart. If you were too decent to do so, you were termed a dope, and the dopest of them all was termed 'the academy dope'. BBS is only a cliché to convey the sense of managing your affairs yourself. Stealing has no place in the academy where you are grooming officers, who by definition are gentlemen as well.

The academy gave a cultural shock to the new comers. The new entrants, joining as first termers, dreamt of joining the best institution in the world and becoming officers. Some were already imagining a twinkling star on their shoulders. The first sixty minutes in the academy brought their dreams crashing down to mother earth. The first thing on the menu was a shaven head and a shaven face, never mind whether you had the semblance of a moustache or not. Most did not at that stage. This left the cadets with an unwelcome new identity. The mothers actually failed to recognize their sons when they went home after the term break. The NDA nuts have created such a collection of abuses that would put to shame the US marines at their best, and turn Shakespeare in his grave. Initially the abuses hurt harder than the hardest punches, and they came at rapid rate. Soon you learnt to hear without listening, ignore and survive. In from one ear and out from the other. Ask any ex NDA, he would rattle them out faster than you can your name.

Well back to the pristine NDA. The rolling hills around, the wonderful Khadakwasla lake surrounded by wooded hills with clean and still waters, looking like a lagoon is a sight to behold. The identically made fifteen squadrons of red stone, the awesome Cadets' mess that can feed 2000 cadets with a five course meal in less than half an hour and the beautifully laid out roads with the majestic Ashoka Pillar is a fascinating sight. Standing atop Point 2475, you get the perspective of the length and breadth of the academy. In that vast expanse stands out the majestic Sudan Block, visible from any corner of the academy. It has an academic cum office block, which has an imposing dome overlooking the academy like the silent sentinel. Somewhere there on the first floor, overlooking a large part of the academy, sits the Academy Adjutant - the hand that rocks the cradle, often violently.

Located in front of the Sudan lawns is the majestic Ashoka Pillar, from where emanate four roads in four directions. In NDA, this Ashoka Pillar, is called the *"Academy D**k"*. It signifies that when the Academy Cadet Adjutant (ACA) finds the discipline of the academy needs tightening, the screw tightening starts from here. It was the centre of mass disciplining whenever the ACA chose to amuse himself, normally on a Saturday, just to ensure that your weekend starts on a front rolling note. The ACA, the show master, stands majestically and monitors the circus from Ashoka pillar, deriving immense pleasure and satisfaction, while his ruthless sergeants literally skin the cadets' knees. So, I guess, there could not be a more symbolic name for the Ashoka Pillar. And guess what, it has been painted in red, could not have been more symbolic - the epicenter of academy tremors. It is here that the cycles spend maximum time off the ground and on the cadets' heads. Ask any ex NDA and he will remember this with great nostalgia.

The NDA has civilian academic instructors who have served there for decades. They command great respect because the military instructors there have all been their students. It has great culture and traditions that never seem to change, let alone die.

The unending Khetrapal drill square, named after the gallant Second Lieutenant Arun Khetrapal, Param Vir Chakra (Posthumous) of Indo – Pak War 1971, is 450 metres in length and 150 metres in width, where a cadet does less drill and more physical training. It leaves an everlasting memory of his three years on the drill square. For this is where he learns the first few steps of drill, the bedrock of discipline and discipline is the foundation of a good army. He passes out from this ground on the notes of 'Auld Lang Syne' watched by his proud parents. The memories of smartly turned out cadets in Blue and White Patrol uniforms marching superbly to the tune of 'Sare Jahan Se Achha Hindustan Hamara' and the unforgettable 'Colonel Bogey'. These cords become part of his heart, mind and soul forever.

A Science block that brings back the memory of science re-tests. It was the scene of struggle with mathematics and physics. Mr. Nair, a Khalsa, yes a Tambi Khalsa was the sultan of the Physics laboratory. In that laboratory, the Bunsen burner refused to burn brighter, no matter how much you cursed or how hard you banged it on the table. The mercury in the thermometer refused to rise as if in a well orchestrated joint conspiracy. When the clock seemed to be breaking all speed records, and with cadets in a tearing hurry, Mr. Nair, seemed to be in no hurry at all to authenticate your reading. For the third battalion types, it was the venue of morning PT during monsoon. It was a small relief, as the area available to make the cadets run around aimlessly was comparatively smaller than the main PT grounds and fewer vertical ropes to hang them by.

A stroll in the Equitation Lines is a horse lover's delight. Close to two hundred odd solid horses with shining coats in different shades, over fed and under worked. And the first term cadets, who are supposed to learn riding, are under fed and over worked, courtesy the second termers, not that there was any shortage of food.

The imposing Equitation Lines is where a cadet runs more than the horse, where he sees the horse's belly more than he

NDA riding class – Who rides whom?

sees his back. And when he does get the rare chance to get on the horse's back, it is crossed stirrups and hours of trotting that bruise your thighs raw, literally giving you a pain in the arse that lasts long enough for next riding class to arrive. And the cycle goes on. After one riding class of three periods, a cadet's walk is worse than a nine months old pregnant woman. While riding would be fun that many would love to do, it is a punishment for most cadets, possibly because of the manner in which it is taught. This notwithstanding, many cadets have gone on to represent India and won medals in international events.

Well, it is not all fun and games for the horse either. He watches in utter bewilderment as the cadets scurry around him in circles and crawl under his belly from side to side. All his training has not prepared him to handle the animal called 'first termer', who in the best of times is a danger to own troops.

The horses' problems get compounded when the first termer mounts the horse, which is rare. For every move that he is trained to execute on a simple command, he gets a dozen conflicting commands. If all that fails to move the thoroughly confused horse, he will dig in both heels and pull the reins with all his might, threatening to tear the horse's jaws. It is not rare that the horse rears up and simply falls backwards. And when none of that makes sense to the horse despite his horse sense, he would whip the horse on the neck, further confusing the hell out of the poor animal. Yet, most are hoping like hell that the horse would stand still for the rest of the class. Suddenly, some horses in utter disgust would take the law into their own legs and take off to the stables at full gallop. The first termer then is at the horse's mercy, hanging from the neck with one hand and grabbing his pith hat with the other, remembering God. Invariably the horse makes it to the stable alone. The first termer, if lucky to have all the bones intact, takes the long walk back to the stables with a gait Charlie Chaplin would find hard to copy. If the first termer does manage to reach the stables with the horse, it is mostly under slung from the horse's neck.

Interestingly, proving the theory of evolution, the horses have learnt to manage the *"Confusionist first termer"*. They have learnt to understand the verbal command of the riding instructor, like *"right forward trot, left foot forward canter or right / left turn halt etc,"* and without waiting for the command from the rider, they simply take off as per the instructor's command.

And believe me, like the cadets, they also have learnt the art of shamming. Their quickest response comes on the command *"right/left turn, halt"* and *"break off, move to the stables".* They stop dead in their tracks, often throwing the riders to kiss the mother earth, even before the rider can say Jack Robinson.

Yeah, I forgot the swimming pool. Majority of the cadets coming to NDA are non swimmers. With no idea of swimming, crossing fifty metres of pool length seems impossible. That was not all, you had to jump from a ten metre board, quite literally waterloo for some. The ten metre board looks innocent from below. It is a different story from up there. For those perched up there, everyone on the ground looks like a Lilliputian and you can see the whole academy, which gives you an 'elevated sense of misery'. While the shortest route out of that misery is the ten metres straight down, many take six months more and many a miles to travel those ten metres.

We were all excited about swimming. In the first few periods, it was a case of so near yet so far. We never touched water. We were put through what they call 'dry swimming'. It involved lying on our stomach on the pool side tarmac and practice swimming movements of hands and legs till the instructor was satisfied. Then you were ordered to jump into the pool.

Only then you realized that all the dry swimming training on the tarmac sinks without a trace in the pool.

Cadet Jagbir Kalia in my squadron was a non swimmer and was working real hard at the dry swimming. I guess he worked too hard, for, at the end of dry swimming, he seemed to have developed a manufacturing defect. He could never get himself to float horizontally in water. He found a novel way, learnt to swim vertically. He passed his swimming test in vertical breast stroke position, popping in and out of water like a well oiled piston. As he kicked and puffed, he sprayed water through his mouth every time he popped out of water, accompanied by a whistle. His forward displacement was so slow, it would make a snail feel that it was over speeding. By the time Kalia finished his fifty metres test, the whole course had finished their breakfast break. Full points to Kalia for his determination. Someone suggested that he try for a place in the Guinness book of world record for the slowest fifty metres in a most unconventional swimming style.

The cadets join the NDA from a vast variety of background and schools. They come in all hues and colours, shapes and sizes; from convent schools, Sainik and Military Schools, from villages and the metros. The NDA grinds you to bring everyone to the same level, breaks you and then builds you into good officer material who have done the nation proud every time the country's integrity was threatened.

Returning to the academy after passing out, remains the desire of every NDA cadet. Preferably, he would like to come back when the cadets of his time are still there, for his memories are still fresh and would be a hero back in his squadron. He can walk with freedom, not having to sneak around anymore. When he gets married, the desire to bring his wife and show all that he did is even stronger. When he does, he will certainly visit his first term and the sixth term room, if he was an appointment. Have a hard look at the floor which he had scrubbed for hours till he could see his face. Tell his wife how he battled the bed bugs, sometime sleeping on the floor. He would look at the seventh heaven, the wire he hung from, nearly severing his fingers. The bathroom where they bathed en masse, rolled under hot and cold shower in birthday suit, and did countless push-ups with his feet on the central wash basin. A steady gaze at the battalion area to try and locate the bicycle he rode as a cadet. He would remember the tricks he would have resorted to- to preserve the tube valves, an epidemic that sometimes hits NDA. The tube valve epidemic is a mind boggling phenomena that starts when the sergeants flatten cycles for indiscipline and confiscate the valves. Or for some reason, that God only knows, the unending cycle of keeping spare valves for the rainy day takes off. The battalion quadrangle also reminds him of the place where he did PT in sweltering heat, which was less PT and more punishment and the dreaded heat treatment etc.

And of course, the trip would not be complete without a visit to that magnificent cadets' mess which can feed 2000 cadets sitting breakfast in flat fifteen minutes. Deep down in his heart, he wants to sit and have a meal there again in his new avatar, a meal in peace. A drive past the 'Hut of Remembrance', the memorial for hundreds of fallen cadets, who have sacrificed their lives in the service of the nation, is mandatory. The hut was constructed by the cadets themselves for their fallen comrades in 1957.

A drive around the academy has to be part of the itinerary. The third battalion types would drive around the Periphery road, pause and look at Point 2475, wondering the number of times they had run up and down this hill, they could have climbed Mount Everest a dozen times. And then they wonder—how could they actually do all that they did. Remembering, it was the sadist fourth term Corporal who made it possible, by the fear he instilled by his bark, which was worse than his bite. Finally, he would stand at the drill square, remember his days with the deadly drill instructors and relive that wonderful tune of the academy band the *'Colonel Bogey'* and the *'Auld Lang Syne'* on which he passed out past the quarter deck in that elegant slow march. Somewhere, deep in his heart, he longs to live that life again.

So these were the landmarks in the academy. The NDA taught the cadets to take tough times in their stride, without complaining. From first term to the last, there was just one quote on your study table, and that was *"Cribbing is un-officer like"*. Well, one could argue against that, because as an officer, you are expected to report a wrong doing by a junior or a senior. Not doing so is a military offence. Here the concept was pretty simple, to groom the cadets to take the rough and tough of life with a smile. I have little doubt that it helped them cope with tough times in the army, later in their career.

Well, personally I support the academy traditions. It taught you many things and above all, esprit de corps and camaraderie, the foundations of the army. One of the reasons for great NDA course spirit is the tough times from mass punishments, for wrongs done by one, known to all, but refusal to reveal the one responsible. A cadet would never complain against a colleague whatever the consequences. If a colleague doped, the whole course went through a severe ragging but never gave out the defaulter's name.

This spirit of camaraderie was best demonstrated in this incident.

Sometime around 1977/78, a fourth term class was waiting for a civilian instructor to arrive. He was late and this instructor was a painful character the cadets despised. A cadet walked up to the blackboard and drew an ass on the board. He then wrote the instructor's name on it, followed by a flier *"..... is an ass"* or words to that effect.

The class had a good laugh and the matter was forgotten. Also forgotten was the sketch that remained on the board. Shortly, the instructor walked in, and seeing the board, went

into a rage.
"Who has made this?" he demanded. No one owned up.
"I say who has made this?" he shouted, trembling with rage. "Someone own up or the whole class will repent."

No one owned up and no one said who had done it. The instructor then had a brain wave. He asked everyone to show their note books. Having compared the handwriting, he unilaterally concluded that Cadet Ravinder Singh was the culprit who had written it.

He had the class fall out outside the room and reported the matter to the academy adjutant. Being a very senior instructor, the matter was taken seriously. The matter was reported to the commandant. After some investigations, the powers that be, decided to relegate the cadet. Relegation meant losing six months seniority, being brought to the next junior term and going through the grind for additional six months. That is the worst thing that can happen to a cadet.

The day soon arrived. Cadet Ravinder Singh stood outside the commandant's office waiting to be marched up to him. The two buglers whose bugle calls announce the relegation, stood outside the commandant's office waiting on the starting blocks, a ritual they are not unfamiliar with. The adjutant came out of the commandant's office, briefed the Subedar Major and all was ready to set the orderly room procedure in motion. The Subedar Major took post as the adjutant marched into the commandant's office.

"March in the defaulter," the commandant said. The cadet was marched in, his stiff starched uniforms rustled as his arms brushed his shorts in perfect unison.

"Will you tell me why did you insult your instructor?" asked the commandant.

"I did not do that sir," replied Cadet Ravinder.

"You are telling lies cadet. The instructor compared your writing with the one on the board," said the commandant.

"I have not done it, sir. The instructor has got it wrong," replied the cadet.

"It is a serious offence, I am not going to have cadets insulting my instructors."

"But I did not do it, sir," remarked the cadet again.

"Were you in the class, when it happened?"

"Yes sir, I was," replied the cadet.

"So you know who did it."

"Yes sir," replied the cadet.

"So, as the matter is indeed serious, either you got to tell me who did it, or else you face the music," warned the commandant.

"I am sorry sir, with due respect, I have been taught in this academy from day one not to ever complain against my colleagues. This tradition has been passed down the ages. So while I know, I shall not give his name. It is up to that cadet to own up and walk up to you, sir." The commandant was annoyed and impressed at the same time. He appreciated that Cadet Ravinder was risking relegation for the sake of his colleague, despite apparently being innocent. But he had made up his mind. The trend had to be nipped in the bud.

The buglers blew the bugle. The sound echoed in the corridors of Sudan Block and reverberated through out the academy, as if protesting against the verdict. Everyone knew that Cadet Ravinder was at the wrong end of military justice, sympathized with him and appreciated his indomitable spirit.

As the bugle sounded, the Subedar Major walked up to him and removed the fourth term badge from his collar. Cadet Ravinder was relegated, the worst punishment in the academy that affects your whole career. Pricked by his conscience, Cadet Ram Prakash had a terrible week. Unable to stomach the guilt any longer, he walked up to the squadron commander and confessed that he was the defaulter and that Cadet Ravinder was wrongly punished.

The matter was reported to the commandant. I am told the relegation of Cadet Ravinder Singh was rescinded. Cadet Ram Prakash was awarded 28 days restrictions. Both of them passed out with their course.

Well, that, then this is the spirit of the NDA. That is the spirit that wins you the likes of the Kargil War. When your life hangs by a slender thread in your buddy's hands and you know he will not let you down, even at the cost of his own, it makes a great army and ours is one. It is this 'Great Spirit' that tides the impossible. It is this stuff that our young officers are made up of.

The Academy As It Was

In this glorious place, the life of a cadet starts in the first term. The term lasts an unending six months, a period, he is convinced will never end and dreams if he will ever get to the sixth term, where he will not have to cross his satchel, walk instead of run, eat breakfast in peace, go to bathroom without a gown, and be an emperor of all he surveys. Now one wonders, how such small things, that mean nothing now, can mean so much at some point of time in life. A sixth termer is no more than 19 or 20 years of age. He looks so mature, so awesome, so much larger than life and unreachable. And when he gets to the unit as a Lieutenant, he looks so vulnerable, so ordinary and so insignificant.

A first termer, is the scum of the earth, a worm, a mistaken birth, an unfit human being and a clot of the highest order, who has no business to disgrace the wonderful NDA by his presence. This is how even a second termer sees a first termer, who himself was in the same boat only a few weeks earlier. A first termer feels like an inmate of a Nazi concentration camp who can be executed without a thought with no questions asked and no explanations given, and no hope of ever getting through the term, quite like the inmates of the holocaust.

At every stage, he is expecting a gas chamber where he will be snuffed. Every day spent is a milestone and counts the DLTGH, an account of "Days Left To Go Home." Nowhere else on earth do the clock arms turn slower and the heart beats faster. I am told that NDA is a changed place now, a lot more cadet friendly. I guess they may not have the same stories to tell as the ex NDAs of the yester years. These stories will give them some idea of the good old days.

My first day at NDA. It was a very long time ago, June 1975, when I drove into Lima Squadron, but I remember that day very well. I was a late arrival and the term had commenced a fortnight earlier. It was raining heavily as I drove in an auto and stopped outside the porch of Lima squadron. The squadron had fallen in for the games parade. The fourth termers with hockey sticks in hand were dribbling the ball in the porch. As my auto stopped, the ball rolled off the porch in the rain. No one moved to get the ball. I was sporting long hair with two top buttons of the shirt open. As I stepped out of the auto, one fourth termer walked up to me and asked if there was anything he could do and if there was any one I wanted to meet. Then he glanced at my steel trunk and bedding in the auto and asked if I had come to join the academy. As I said 'Yah', he said, "Not yah, you son of a gun. Say yes sir and go and get that ball ok." I walked out in the pouring rain and retrieved the ball, getting the first taste of things to come, a bitter taste indeed!

There was sudden excitement all around as if the hungry big cats had just found a new prey. They were quite surprised to see the way I had dressed. Those were the days of long hair and I had left my school long ago. I was taken to the Squadron commander who was equally appalled to see me. He asked the Cadet Sergeant Major why I was not wearing a tie. My next stop was the barber shop. The fourth term corporal who accompanied me gave some instructions to the barber and the ten minutes there changed my identity. I could not recognize myself. From flowing shoulder length hair to absolute military zero cut. That was the way it would stay for many years.

I was shown my room, which had all black painted furniture. A large cupboard cum dressing table with a sizeable mirror, a study table with a study chair, a bed and a rack. Grey tiled floor, on which we would be spending a large part of our time scrubbing to make it shine, completed the picture. All in all a very comfortable arrangement; an independent room for every cadet. A wire mesh about two feet broad all along the front wall above the door caught my attention. It allowed one to peep into the room. I did not realize then that it was going to play a major role in the academy and take us to the painful seventh heaven. From my room I saw the first termers run around in the rain. I was given time to open my box. After the games the filthy first termers proceeded for a bath. In NDA it is 'community bathing in birthday suit'. While I had heard about it, a dozen dangling danglers and bums of all shapes and sizes was a cultural shock. Soon one gets used to this melee and barely notices the august gathering around. The study period was uneventful, except that I also sat at the black study table staring at the wall in front with nothing to study and not allowed to get up. I was happy to be in the NDA, but wondered what lay ahead.

The silence on the ground floor was deafening and scary. The only sound was the ceiling fan rattling with every rotation, a sound that was to stay with me for a long-long time, for that sound was associated with the tortuous first two terms in the academy. That noisy fan would give away my presence in the room, despite a deceptive lock outside with unpleasant results. Soon I learnt to live without it and breathe noiselessly. My first ante-room session was a revelation. All first termers attended the half an hour order session in "Murga position" (Rooster position). Wonder what orders we would have registered in that holy position! The dinner was a case of so near, yet so far. In that impressive dining hall with a tempting multi course dinner, I was told that I could eat as much as I wanted, after I finish my watery soup with a fork in my left hand. The futile exercise with soup was followed by what they called 'Blow hot- blow cold' - alternating swigs of hot coffee and ice cold water. I had the first taste of a first termer's five course dinner.

Then on, like the rest, I was like a dazed robot following the first termers in a "bher chaal" (sheep walk). Everyone responded to the instructions of the barking corporal like over wound toys, huddling to remain in the centre of the flock to keep out of arms length of the ever angry corporal.

Tragicomedy Beyond Comprehension

The NDA gives a cultural shock to many who land up there with great expectations; expectations raised from joining the finest academy in the world, that will launch them into a wonderful career and would become a proud officer in the Indian armed forces. It's quite funny, the public school types landing up with the 'Howdy pal?' kind of public school English soon realize or, more aptly, are kicked into realizing that their survival lies in learning and speaking jack boot NDA English.

One such first termer who landed up at the academy with great expectations and visions of becoming an officer narrated to me his very unusual and quite incredible experience. Beant, the unfortunate gentleman is a dear friend. Here it goes in his own words :

Quote "Yaar, I came to the NDA with great expectations

and hopes, since I had heard so much about it. Those days the first termers used to be housed on the ground floor of the squadron. I was a typically simple highlander from Himachal Pradesh and could not imagine the ways of the academy in the wildest of dreams.

To top it all, *meri kismet kharab* (my luck was bad) and I had the misfortune of being detailed as the understudy to the Cadet Sergeant Major (CSM), a traditionally sadistic key appointment in the squadron. The understudy as you know is more popularly known as the "CSM's undy." Initially I was a little thrilled with myself since I was under the impression that the smartest of the first term dopes was picked as the CSM's understudy. His duty was to assist the CSM in maintaining parade state and recording orders, etc. It did not take me long to realize the folly of my judgment about my smartness.

Having been nominated the undy, the CSM called me, and our conversation proceeded like this:

"You son of your neighbour, do you know what you are supposed to do?"

I was taken aback with the abuses as I had never heard these before, but could do sweet damn all about it.

"Yes sir," I replied.

"Louder you clown."

"Yes sir," I shouted louder.

"You stinking rotten clot, when I tell you to shout, it means shout. I want the damn squadron to know that a clot is talking to the CSM of Alpha squadron, ok?"

"Yes sir," I shouted my guts out, feeling like a clot.

"That's much better. And now listen carefully. Your first duty is to wake me up every day at five o' clock in the morning with a cup of tea and eats. Remember, five o' clock means five o' clock plus minus two seconds and no more. You make a mistake on timing and you will curse the day your mother met your dad, ok?"

I spent the whole day brooding, depressed at the most incredible abuses that I had to endure for no rhyme or reason. All my excitement about being the smartest first termer disappeared without a trace. I spent a sleepless night gazing at the fan which rattled once in a while, rattling me in the process and giving me jitters every time. Hardly sleeping, I spent the whole night thinking about how to comply with the plus- minus two seconds deadline. The consequences of not doing so did not leave much doubt in my mind.

I had taken the CSM's tea mug the previous evening. Unable to sleep well because of the worry, I got up early in the morning, dressed up and proceeded to the tea room to fetch his tea. As it always happens, a first termer getting tea for a senior is hijacked en route only to find himself loaded with ten more mugs, one hanging precariously from each finger. Well it is a case of tea –tea everywhere and not a drop to drink since none of these mugs is his, till he learns the art of survival - drinking from the other's en route. Having managed to dispose of other mugs, I proceeded to wake up His Excellency the CSM, Alpha Squadron. I reached his cabin and waited outside with eyes fixed on my wrist watch. At sharp ten seconds to five, I knocked the door and in the politest of voices that I had rehearsed over and over again the previous night, said:

"Excuse me sir, please may I come in sir?"

"Yes come in, who is it?"

"Good morning sir. I am the understudy. It is five o' clock sir, I have brought your tea sir."

"Come in you clot and keep it on the peg table."

I kept his mug on the table, "Thank you sir, excuse me sir, please may I have your permission to leave sir," I said, wanting not to make any mistake and turned to leave the room. What an irony! I was thanking him for giving me the honour to serve him his bed tea, for which that clot should have been thanking me.

Before I could leave the room he said, "Come here you son

of a gun." My heart skipped a beat as I wondered where had I screwed up now, and I stepped closer.

He said, "Come closer, you crap!" I stepped even closer.

"Come closer you son of a b**ch". I moved even closer and my face was no more than six inches from his as I felt his foul tobacco breath. I stood there wondering what was coming next and before I could figure that out, he registered a stunning dhap (slap) on my face that shook the day light out of me, and that even before he was awake. And for me, I was more than fully awake.

Nature also plays its pranks. The stunning dhap announced the day break as the golden ray of sun light filtered through the blinds of his window. What poetic injustice.

"Get lost now," he said without any remorse.

I was stunned. The day had begun with a bang of the wrong kind. The whole day I wondered where I had blundered, while the stunning slap cast the shadows of the coming events. The next night was spent in figuring out where I had gone wrong, but I could not, getting a headache in the process. I rehearsed the entire thing before going to bed in an effort to ensure that I made no mistake the next day.

So came the next day and, lo and behold, I was there in front of the CSM's cabin five seconds to five o' clock, determined not to make any mistakes this time. I went through exactly the same conversation as the other day, starting from taking permission to enter the room adding more 'Sir's than before, in the hope of not annoying him. Sure enough, events began to unfold themselves in exactly the same fashion. After a series of "come closer you scum", the event, again, closed with a stunning slap.

Exactly the same thing happened over the next three days without any better luck. After that the ritual would proceed something like this, except that there were no dialogues exchanged. I would go with his tea, knock the door and say :

"Good morning sir, please may I come in sir. I have brought your tea sir."

"Yes come in, you crap," he would reply in a sleepy voice.

No more words were exchanged. I would enter the cabin, keep his tea mug on the bed side peg table, and as a programmed robot move closer to him and stretch my neck towards His Excellency, the CSM. He would then register a dhap on my face without pity, without remorse and without a thought. All this like a well rehearsed light and sound programme. I would thank him and leave his room.

Can you believe, it carried on that way without a break for full six months of the first term. Can you imagine starting every single day with a slap for six months in this fashion. I thought I was here to become an officer. Unquote.

NDA has a fine tradition of 'End of term social'. On this day, a few days before the passing out parade, the sixth termers place themselves at the mercy of the squadron cadets. The squadron then pays them back in the same coin. Those who had been sadists are returned favours with interest, apart from serving tea to the first termers, etc. The gentleman variety gets through without too much trouble. As for the CSM, he has a busy day on the social, for his job demands that he be a tough ruthless cookie to keep a hundred ticking brains under control. Ever since a tragic incident occurred during the social, in the 48th NDA course, the tradition has been diluted a great deal. Nevertheless, the tradition was a constant reminder to the seniors that every dog has his day. And above all, everything was forgotten and everyone parted company as good friends after the social, leaving unpleasant memories behind for good. Beant unfortunately could not lay his hands on him, for others were too occupied with him settling their scores.

"Ever since then it has been my ambition in life to find him and ask him whether there was something wrong with him or me. Unfortunately I have never found him," said Beant.

It may seem more sadistic than funny, but imagining the event happening sequentially day after day in total silence was as much funny as it was sadistic. As for the great CSM, no big deal, because such characters are in abundance in the NDA.

No one would, I am sure, fancy being in the position in which my friend found himself. This, in brief, sums up the life of a first termer those days. Interestingly, the more panga you had with a senior, the better friends you became by the time you left NDA.

As I said, the first termer was the lowest form of living being on earth.

Lele vs Pethe

Well, Sundays were worse than other days for the first termers. He was at the mercy of the wolves the whole day if he got stuck with anyone early on a Sunday. He could be doing errands for seniors throughout the day and some of them could be really silly. Smoking was banned in the academy. Some cadets smoked nevertheless, and the smokers' chain could have a link in all the different battalions. A first termer would be caught on a Sunday and sent by Cadet X in first battalion to Cadet Y in second battalion to get a fag (cigarette). Cadet Y could then send him to Cadet Z in the third battalion to get a fag from him for himself. Cadet Z would keep him doing bend stretches till he found a cigarette hidden in the curtain rod. The first termer takes the long walk back to Cadet Y with two cigarettes ensuring they remained in good shape, all the while being extremely vulnerable to being dragged into any squadron and given some treatment for no reason except being a first termer within reach. Cadet Y would keep one cigarette and send the other to Cadet X in the first battalion. A few hours and many a mile would have been traversed for the lone cigarette. When he finally gets back, he would most certainly be kicked for getting late. Sadly, much to his consternation, his day has just begun as someone is already lying in ambush for him for another duty, which could be bringing two rupees worth of chikkies or jalebis from the Gole market. There he would find many of his ilk on similar errands. He would be handed over to someone else when he gets to the squadron. So the cycle goes on through the day.

Many first termers would skip that wonderful Sunday breakfast and disappear early morning, hiding in variety of places like mandir, school/place type's cabin or Gole market etc. Some, even if caught, would take that risk and desert en route, prepared to accept the consequences at the end of the day, rather than spoil the whole day.

In my squadron we had a long array of tants, which in the academy parlance meant Maharashtrians. They acquired this sobriquet from the famous Maratha warrior, Tantya Tope. This list of modern day warriors included Donde, Thatte, Sathe, Bhonde, Pethe and Lele and to close the door on them was Phatak, pronounced as Phataak. This list when rattled at 90 RPM, made a good rhythmic poem. Of these, Lele was our fourth term corporal, five feet f-all, a bundle of energy entirely focused on making us miserable. Pethe, my course mate was a six footer well built bloke. One Saturday, Lele asked him to report the next day morning for some duty. Pethe realized that reporting for one duty meant that he was stuck for the day. He said, "To hell with him, I will not go, will give him some pill in the evening. Arre yaar, he is my tant type, he won't say anything." Sunday was spent in peace hiding somewhere. Come evening and we were back in the squadron for the evening fall in. There was Corporal Lele in the squadron centre lobby, waiting impatiently for Pethe. The moment Pethe landed in the squadron, Lele grabbed him by the tie and their rather interesting drama proceeded like this:

"Bastard, I told you to report to me in the morning, why didn't you report to me?" asked Lele red with rage, holding Pethe's tie well above his own head.

"Sir I was called...." Pethe began his well rehearsed reply. But before he could complete, Lele jumped and slapped Pethe hard, shouting, "Shut up, you bastard," and continued;

"Bloody clot, I told you to report to me in the morning, tell me why the hell didn't you come? Answer me you idiot, yap out," asked Lele as he shook Pethe up and down with his tie, rising on his toes to see Pethe in the eye.

"Sir I was ...." Pethe tried to reply. But as before, Lele dhaaped Pethe, shouting, "Shut up, you clown, feel like talking. Don't bloody yap."

Lele pulled Pethe to the staircase and stood on the second step. From here, he was at par with Pethe, breathing into his face, eyes popping out.

The interrogation continued for next half an hour without getting anywhere. Lele kept asking him why he had not reported and as Pethe would begin to answer, Lele would dhaap him and tell him to shut up. Hell'va dilemma Pethe had found himself in- being asked a question repeatedly and slapped and told to shut up when he did. Another case of 'tragicomedy'. The tant type net did not work unfortunately. He was rescued briefly by the arrival of the Squadron CSM. His misery would continue throughout the night though. It was well worth having the Sunday to you, even if the night was not.

Howl You Clot or.. !

Coming from a civilian background, most cadets take time to understand the NDA lingo. The first few days are a shock treatment. The first termer's life revolves around two things. First, getting hold of one square meal in a day, by hook or crook, and the second, getting through the day, without too many bruises. Both, no mean tasks.

Getting to the mess was like the last two hundred yards to the objective-full of mine fields. You could get blown up anytime depending on who the sergeant was on post. For the third battalion types, it was less than two hundred yards, which often proved far too long. The cadets in the NDA can move from one place to another only in a squad of four or six, whether on foot or on bicycles.

Cadet Rehal, a simple rustic surd from village Urmur Tanda in Punjab, was grappling to come to terms with what he was confronted with in the new place he had found himself in. Not lucky enough to get breakfast that day, he was dying to get to the mess for lunch and hoping he would get to eat. Rehal joined a squad and marched to the mess. Not wanting to attract the attention of the sergeant on post, he swung and dug wildly, as if his life depended on that, (I guess it did), his hands threatening to knock off the turban of the cadet in front of him and teeth of the one behind.

The sergeant watched in amusement, as Rehal, popping up and down marched like an erratic robot. The squad marched past the sergeant holding their breath, relieved that they had made it.

"Squad halt, about turn," roared the sergeant. Their heart sank. Less than fifty metres to go, they were marching away from the mess. The squad turned around and halted in front of the sergeant.

"You bloody khalsa, where do you think you are, and what

do you think you are doing, drill or bloody bhangra (a lively Punjabi folk dance)? Which place do you disgrace?" asked the sergeant.

"Sir, Urmur Tanda."

"Where the hell on earth is this funny place?"

"Sir, Urmur Tanda," repeated Rehal meekly.

"Louder, you monkey," shouted the sergeant, getting visibly irritated.

Rehal spoke still louder, but the sergeant was not happy at all.

"Louder, you stinking clot," yelled the sergeant.

Rehal shouted his guts out but the sergeant was still not impressed.

"You son of a gun, piddly first termer, when I say louder you better howl, you understand- howl before I bury you right here, you get it?"

"Yes sir," said Rehal, not quite getting it. He stood there confused and frozen, not quite knowing what he was expected to do and then.

"Eeeeeee-yaahoooo-ooooooh," howled Rehal like a jackal.

The squad burst out laughing. "Shut your stinking traps, you idiots," shouted the sergeant, taking the squad by surprise. The laughter disappeared in a flash, faster than it had arrived. Smiles, however, still remained.

"Wipe the smile off your faces," said the sergeant. The cadets, as if in a drill square unison, lifted their right hand and wiped the smile off their faces, clinching their fists as if holding it in their hands.

"Now shove it up your backside," continued the sergeant.

In a drill square like precision, the squad swung their right hand behind their butts and did so. The sergeant, then surprising everyone, smiled, "Turn right, march off." They were heading back towards the mess for lunch without much damage, Rehal taking no chances as he continued to dig and swing dangerously.

That day Rehal learnt, howl in NDA means to shout your guts out and not howl like a jackal. Rehal's rise to fame in the academy had just begun.

Thou Shall Not Steal, Only Manage

Managing or colloquially called BBS was the simple mantra by which the NDA cadets learned to live and survive day to day. Contrary to what this may suggest, no one steals in the NDA. It is strictly un-officer like and simply not done and never tolerated. Everyone except the first termers have enough juniors at their beck and call, so never a problem. As for the first termers, they just beg or borrow. In other words-manage. Well, in the NDA, it is a survival and management technique and teaches you a few tricks for winning wars – builds an attitude to 'find a way or make one' but never complain. There simply is no excuse for not doing what ought to be done, in the manner it ought to be done, in the dress it ought to be done and at the time and place it ought to be done. In war a soldier will have to live off the land, survive on strong will power and thin air, innovate and win no matter what. In war, it is the mental robustness developed in the academy that makes the difference.

Somehow, the cadets keep losing their kit, one way or the other, yet everyone seems to have everything when required. I strongly believe no one has ever paid for any lost piece of equipment or clothing. Only, I wonder what would happen to the last course, if the academy was to wind up one day.

Well, let me tell you how the whole cycle of BBS or managing is set into motion. I joined Lima Squadron with the 54 NDA Course, way back in Jun 1975. Soon we were issued our clothing. I remember a second termer, Cadet Bahri (actual name), who had the reputation of being a decent fellow had been pretty friendly with me during the month that we had been in the academy. Friendship between a first and second termer was rather rare, particularly in the same squadron, as they say, "What happens to a cow if she makes friends with grass?" One fine day friend Bahri walked up to me and said,

"Yaar, we have a 'cabin cupboard' (inspection of the living room), by the sergeant. Can I borrow two pairs of OG woolen socks for display. I shall return them tomorrow after the inspection."

In the cabin cupboard, a cadet was required to display all the items of clothing properly laid out. Everything had to be brand new, properly folded with razor sharp edges. So everything from shorts to handkerchiefs and socks had a cardboard inside. Well, coming back to Bahri. Many cadets borrowed things for the cabin cupboard and Bahri was a decent guy. He had been pretty friendly as I said, so, I trusted him and thought I must help him when he needed it.

His cabin cupboard, apparently came and went by. I am not even sure if there was one. Many days went by, and I like a gentleman, waited for him to return the OG socks. When it did not happen for another few days, I walked up to him and asked him, "Sir, now that your cabin cupboard is over, may I have my OG socks back." He looked up and without blinking an eyelid said, "Which socks? I never took any socks from you." I looked at him for a while, a bit stunned. I thought he was a decent guy. I had little choice, I walked away. I learnt many a lessons that day.

Well, I started with two pairs of brand new socks down, even before opening my account. That was the last thing he ever got out of any of my course mates, for I let that be known to all. But then, to be fair to Bahri, I did not hold that against him. I am sure someone had done the same to him somewhere.

The NDA worked like that. But what is likely to happen is that it would set the cycle of BBS in motion. Some argue, such an upbringing can form bad habits for life. Well, I am pretty certain cadets leave those tricks behind as they leave the academy. Nevertheless, no other service comes close to us in honesty and selfless commitment to the nation. We have proved that time and again.

Well back to BBS. The Academy Cadet Adjutant (ACA) noticed something that he knew for years. The cadets were bringing their beret caps inside the cadets' mess. He passed directions that hence-forth no beret caps will be brought inside the mess. They will be left outside on the hangers meant for the purpose. My guess is that the instructions came from some officer who wanted to introduce the same customs followed in the unit Officers' Messes in the army. He obviously failed to appreciate its implications in the NDA.

I have no doubt the cadets saw the writing on the hangers. No one obeyed the ACA's orders. There were no caps on the hangers and yet, no cadet seemed to be having one in his hand in the mess. Where were the caps, the sergeants wondered. The sergeants took post at the entrance of the mess and forced the cadets to hang their berets on the hangers. So against all their wishes, the cadets were forced to leave their berets outside on the huge circular steel hangers provided for the purpose, but never used so far. The first day was uneventful; the second disastrous.

The next day, the cadets on return from lunch found their new berets missing and the old ragged ones hanging in their place. It was quite clear that the berets that were left were specially brought to the mess for the purpose of swapping with the new ones. Using such old ones was clearly an invitation for trouble, no sensible cadet would ever risk. Always a battle of wits, the cadets found a way to save their berets from the crooks. Before entering the mess they would go to the toilet and tuck their berets inside the shorts and carry them inside. It succeeded for sometime.

The sergeants soon noticed the decreasing numbers of berets on the racks. Always known to be bigger crooks, the sergeants soon discovered the secret of missing berets. The bulging shorts gave them away. The first unlucky ones to be caught were slaughtered. The rest learned fast enough and the hangers were once again brimming with berets. History repeated itself and the good berets started to vanish again. Some tried to beat the crooks by bringing old berets for lunch. This also had its attendant risks. If you were caught by a sergeant with an old beret, you could be hanged by your b**ls and it could well might ruin the next few days and many more, not to mention the confiscated beret. So, as said by a wise man, "Necessity is the mother of invention", the cadets came up with a brilliant idea, and that was to lock the berets to the steel hanger with a steel lock and chain through a hole in the beret. The locks and chain were otherwise used for locking the bicycles in the academy. Ever heard of such brilliance, and people associate the military with brawns rather than brains. Funny, isn't' it? The battle of wits had just begun.

Well the crooks couldn't digest that. It did not take them long to come up with a counter measure. And when the crooks struck across the academy, the dopes found their steel locks and chains dangling from the hangers and the berets gone. The crooks had simply cut the berets off with a blade leaving the locks and chains. It was a funny sight, their expressions, one of disbelief, unlocking their chains and then taking the long walk back to the squadrons without caps. And it seemed half the academy was walking back bare headed. The crooks had a field day. They seemed to have built up enough reserves for the next generation. The cadets, always up to mischief, mostly did that for fun.

That, then, is the story of the survival of the fittest in the

NDA. Well the ACA, or whoever gave that order, finally saw the reason and withdrew it, much to the relief of the cadets.

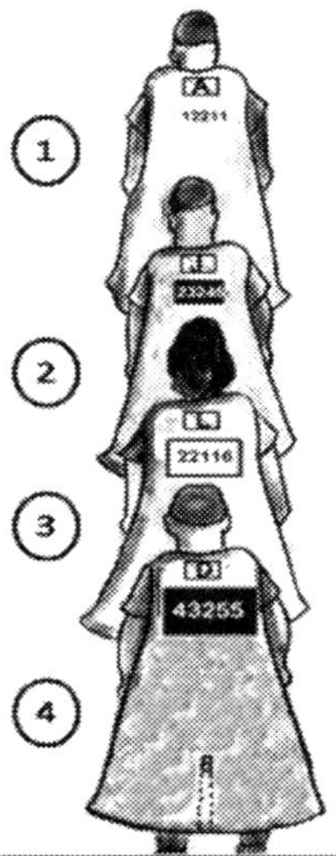

1.	**Initial issue**
2.	**Whacked once**
3.	**Whacked twice**
4.	**Whacked Thrice**

Story of NDA Rain cape

Ek Ke Do, Ek Ke Do

• ***Div O*** *- Divisional Officer, instructor of the rank of Captain equivalent from Army, Navy or Air Force posted in the squadron in NDA. Called 'Div O' for short.*

The academy pays great stress on drill, for drill, as they say in the army, is the bedrock of discipline, and discipline is the foundation of an effective army.

So, in the NDA, the cadets are required to pass the drill test. The incentive for passing drill test is – the privilege to wear a lanyard while the failures wear an arm band on right arm with their academy number on it, and the more valued-the privilege to go to town on Sundays called liberty.

Cadet Shoorveer was a courageous bloke. His name meant exactly that. He had not passed the drill test till his third term. Quite fed up of having to spend his weekends in the academy, while his course mates taunted him and went to town, he decided to proceed on French Liberty (unauthorized trip to town).

Dressed in civvies, he sneaked out and headed to where he thought most cadets would see him, the popular picture hall. A NDA cadet cannot hide himself on the street. He simply walks different, behaves different, looks different and his haircut is a certain give away. As anticipated, he found many of his course mates near the ticketing window waiting to see the premiere show of a much touted film. Much backslapping and suddenly a cadet said, watch out Shoorveer, the Div O is coming. The Div O was walking straight towards them, looking sideways. There was not much time to do the disappearing act. Shoorveer thought like lightening, grabbed the towel of a waiter passing by, covered his head, snatched the film tickets of his course mates and started *"Ek ke Do, Ek Do, sirf do ticket baki, jisko chahie, le lo"* (One ticket for two, last two tickets left, any one interested). Saying this, he began to sneak out.

As he watched from the corner of his eyes he saw the Div O heading straight for him. The Div O has identified him, he thought, surely, he was done. That meant fourteen days restrictions. The Div O stopped next to him and asked, *"Doh ticket hain?"* (Got two tickets?) looking elsewhere, obviously not wanting to be seen buying tickets in black with cadets around. *"Hain saab,"* (Yes, I have) replied Shoorveer head well covered, looking down.

"Kitne rupaye?" (How much?)

"Ek ka doh, doh ticket ke doh saw rupaye saab." (One for two, two hundred rupees for two tickets).

The Div O quickly paid the money and disappeared in haste. In that overcrowded and noisy lobby, the trick worked.

Shoorveer had lived up to his name; he made two hundred bucks from his Div O, tricked him at point blank range and had a free outing. As for his course mates, tough luck mates. With their tickets gone, they will have to do something else for the day. Fortune favours the brave, as they say, don't you agree? Well, the line between bravery and fool hardiness can be very thin but nothing succeeds like success. Shoorveer had become an instant hero.

The Ball and Dummy Exercise

Weapon training is taught to all cadets in the fourth term in the academy. Young kids out of the school, into the frying pan, they were quite excited about it. Mostly they were danger to own troops on the range. One exercise used to improve the firing standards of the cadets was called 'Ball and Dummy exercise'. The exercise is meant to correct flinching by the firer when the shot is fired. Unknown to the firer, it involves putting a dummy round at random in the magazine along with the live rounds. The firer fires and continues to flinch with the bang, but when it is the dummy round in the chamber, there is no fire and no bang but the firer flinches and bucks out of reflex. Thus his mistake is easily noticed and corrected.

Bayonet fighting is also taught to the cadets at the NDA. It aims at instilling an aggressive spirit in them and trains them in hand to hand combat. For bayonet training a stuffed dummy soldier is made. A pole sticks out from the waist depicting his rifle. Cadets practice bayoneting the dummy in various attack modes. It is a great sight as naked aggression is accompanied by a full throated battle cry.

At the end of training, the cadets were put through a

written test. One question required them to write a short note on `Ball and Dummy' exercise.

One wise cadet wrote: *"In bayonet fighting when you bayonet the dummy on its balls, it is called 'Ball and Dummy' exercise"*. The question carried three marks, the officer gave him four. Now, just imagine a cadet backed by a full throated battle cry jumping four feet in the air, bayoneting a dummy on its balls with blood shot eyes, wanting to rip them apart. Well, the dummy does not have any, it is not meant to, since bayonet fighting is not a precision guided hand to hand combat tool. In any case, it is meant to kill the enemy and not castrate him.

Shoot to Clean

For cadets, small arms firing at the range was fun and a much awaited event. But everyone dreaded the weapon cleaning that followed. Hours of hot water pouring and pulling through was treacherous. The kote (armoury) NCO did not help matters by being over strict, to make his own life easy.

The fertile minds of NDA cadets, normally under employed, discussed more efficient ways of cleaning a rifle barrel, rather than waste precious hours pulling a 'pull through chord' through the barrel. A cadet when asked the reason why a `chindi' (piece of soft cloth four by one inch in size, used for cleaning the barrel with a specially designed pull through chord) was placed under the first round in the magazine, answered, "So that when you fire the first round, the 'chindi' also goes with the bullet and cleans the barrel for further firing". It is actually kept to prevent dust entering the magazine and to know if the round falls off from the magazine.

When it was first suggested that the earth was round, the man was beheaded, I am told. The NDA cadet suggested this concept of automated barrel cleaning long ago and some had laughed then. And now you have such rounds which are fired to clean the barrel, saving precious man hours for something more productive - digging and filling trenches.

The Fan Shall Not Move

Be it the drill sergeants, or the cadet sergeants, the academy orders are implemented less in letter and more in spirit. I was a victim of lack of common sense which is so common in the academy, both by default and design.

Leaving lights or fan on when not present in the cabin called for punishment. It was in the form of Sinhgarh hike, which was a long trek up the historic Sinhgarh fort on a Sunday. It screwed the whole weekend, Sunday being the much awaited break from the gruelling routine of the academy.

I was awarded one for a moving fan, when not in cabin. The fan was moving in the opposite direction due to the breeze through the open ventilator. All my pleas fell on deaf ears. No explanation was good enough. The fan is not supposed to move when switched off in NDA, I was told. One Sinhgarh hike awarded. No questions asked, none entertained.

What is not Civil and What is not History is Military Geography ....

The NDA rigmarole was too hectic and physically oriented for the cadets to be bothered about academics in the earlier stages. Barring a few who aspired for multi- starred torches and high CGPAs, most were content to get by. Majority survived either by crisis management or by working out guess papers. Arun Marya, CSM 54 Course, Lima Squadron in the esteemed company of Jagbir Kalia of the swimming fame, became more famous than Elizabeth Taylor with a higher credibility rating than the Pune weather forecasting department for the accuracy of their guess papers. They were the most sought after commodity before the exams and soon forgotten till the next; like the God and soldiers after the war.

There was the other category who survived on the mercy of the instructors, taking pity on the poor cadets who spent most of the time licking mother earth. Then there were instructors who either did not know their subject well enough or realized the futility of trying to put anything in the tough NDA nuts. One such instructor was from the Army Education Corps. This corps occasionally has been the subject of disbandment debates in the army, as some feel the army could do without them. I prefer to call him Captain Choudhary who taught the cadets Military Geography and was very proud of the fact that he was the senior most Captain in the Indian Army.

The subject was introduced in the fourth term in the academy and that was our first day. There was excitement in the air as we waited impatiently for the new subject that sounded more military than anything else we had done before. Added to that was the anxiety of having a new teacher whom we had not seen before. Soon we saw our new instructor stride in with handle bar moustache. His military looks made him a bit out of place in an academic class room. He would have better suited leading the Don Quixote's charge of the Light Brigade, or inspired more confidence on the battlefield of Trafalgar with a musket in hand rather than in a class room with a chalk in hand. He settled down, surveyed the class as if he was Lord Nelson, looking for potential trouble makers. Having made an assessment of the gang, he delivered his maiden speech.

"Gentlemen, welcome to this very interesting subject of Military Geography. I am Captain Choudhary, and I am the senior most Captain in the Army. Very soon, I will be the junior most Major in the Army. Now, coming to the subject, I have been in the academy for almost three years and have been teaching this subject. As you can see for yourself, with the experience behind me, I can be of immense benefit to you if you take interest. I am certain you will not only find it interesting but also be much wiser at the end of the class," pausing to collect his thoughts, he continued.

"And now getting on with the subject proper, the definition of Military Geography. Gentlemen anything which is not civil and which is not history is military geography." So that was the benefit we derived from his long experience. Such instructors who did not waste much energy in class were generally quite helpful in day to day life. So we were quite positive that we would sail through the issue of semester exams without problems. Soon the first semester exams arrived. He had hardly taught anything and the cadets could not be bothered. We cornered the instructor on telling us important questions, which we could cram up and tide over the seemingly difficult issue of the examinations. He told us he would do so in the next period. The next period arrived, and sure enough, the instructor true to his word dictated us five important questions, adding, that of these, any three would be asked in the test. That was fair enough, after all we could not expect him to give us everything on the platter, and he too had to do his *'naukri'* as well.

On the day of the test, the cadets walked in much relaxed, as if proceeding on liberty. They were well prepared for the test-thanks to Captain Choudhary. Three known questions guaranteed pass percentage and a cadet needed no more. He had been showered the choicest of praises. Some said he was the best instructor the academy had. The others classified him as a gem who really understood their plight. Some said he was the most humane of human beings. The cadets settled in the class with total nonchalance and confidence. Some even sang *"What's the use of worrying"* instead of the usual *'Auld Lang Syne'.* There was no anxiety at all; there was no reason to be. Surprisingly, some other instructor strode in to serve the paper. Everyone settled down. There wasn't any effort to find seats next to the brighter lot. Everyone looked simply brilliant, brimming with confidence.

The instructor, after the customary warning against unfair means and talking, served the paper. When we saw the paper, to our horror, not one of the five questions appeared on either side of the paper. We turned the paper inside out not believing what we saw, but that changed nothing, and then as if in unison, the whole class looked up exchanging inexplicable looks. The shock took time to clear. But for the tough NDA nuts, most would have suffered a stroke. The academy always taught you to be cool under pressure. The seniors had told us "*Beg, borrow or steal, but sort your problems out yourself.*" That and the principle of mutual support were applied to tide over the crisis. We wrote stories and thanks to the experience of our instructor who corrected the papers, we performed rather well. Later we learnt that on the day of the test, by then Major Choudhary had reported sick and was relaxing at home, when we were cursing the hell out of him. Suddenly all great opinions about him changed, and in a flash he had turned from a 'gem of a man' to a 'clot of chap'. Later, when we confronted him on the issue, he did not consider it necessary to apologize or explain, saying your motto should be "Hope for the best, be prepared for the worst".

In order to make up for the disaster, we convinced him to at least give us good marks in the class assessment, which carried five marks. He promised he would. Later we learnt from reliable sources that the highest marks given to the brightest of the lot was one and a half, most were awarded half. So that then was the most humane of the human beings. God bless him any way.

The Inimitable Mr. Keshwani and his Jawa

The academy in addition to its military curriculum aims at qualifying the cadets as BA/BSc graduates. There are therefore large numbers of civilian instructors who teach academics. Being a great place, with very little to do, for cadets are hardly interested in academics, they stay on and retire from there. All ex NDA officers posted there have been their students. They thus carry a lot of clout.

As would happen anywhere, there is both the popular and the unpopular variety. Mr. Keshwani, who taught Chemistry, was one of the former kind.

No folklore of the academy can be complete without the mention of the legendary Mr. Keshwani. He was a fine gentleman who was as popular as he was full of life. There is an incident about him which has been a part of the academy lore for a long - long time.

In the NDA the cadets are required to pass the drill test which in the academy is called 'Drill Square Test' or drill square for short. Those who pass the drill square test are permitted to wear a lanyard and salute the officers to pay compliments. They also have the much desired privilege of proceeding on outing to Pune city on holidays, called *'liberty'* in the academy parlance. The drill test failures on the other hand are not permitted to wear the lanyard. Instead, they tie an arm band showing their academy number, reminding one of the Jews in the concentration camps of the Second World War. Unlike those who have passed the drill test, they are not permitted to salute but are supposed to pull their arms to pay compliments to the officers. Passing the drill square test, therefore, is an important benchmark for a cadet.

Coming back to Mr. Keshwani, he had a tremendous personality, but physically, he was not particularly handsome.

He was rather dark and balding on an average frame, and wore thick rimmed glasses resting on a prominent sharp nose. His wife on the other hand was a charming fair lady, and their daughter, even prettier and fairer. Mr. Keshwani was one day heard saying:

"Look at the bloody NDA cadets, when I go to the market alone, no one salutes me. When I go with my wife, half the academy salutes me and when I go with my daughter, even the drill test failures salute me."

Another part of the story is that this anecdote has been reproduced on many occasions in the Readers Digest by an ex NDA cadet. Mr. Keshwani quipped "Look at the bloody cadet, making money at my expense time and again. Trust an ex NDA to do that."

Mr. Keshwani during his younger days, I am told, was a dashing young man. He sported a beautiful Jawa motor cycle. He loved his bike so much that if anyone ever asked for his bike, he would reply, "You can take my wife but not my Jawa". He retired from the academy sometime in late eighties leaving a big vacuum.

- *Squady- Squadron Commander of the rank of Major/ Lieutenant Colonel equivalent from any of the three services.*

Mr. Keshwani had a photographic memory. On asking some cadet's name, he would suddenly ask,"Are you related to so and so of the same surname of so and so course?" and then he would say, "he used to sit in the last row, fourth from left and passed out as so and so appointment." We cadets initially used to think he is pulling a fast one. Some cadets then had that checked out and almost in all cases he was simply right.

Mr. Keshwani had a special affiliation with Lima squadron and was their undeclared guardian. He took extra interest in Lima squadron cadets, whether it was their academic performance or ensuring their attendance in the class.

One day he found Cadets AK Singh and Wallia of 54 Lima missing from the class. On enquiring, some cadet said they are sick in quarters (SIQ), a standard excuse when a cadet was bunking a class. Keshwani had seen far too many cadets to believe everything cadets said with supreme confidence. He left the class and rang the Squadron Commander (Squady). The Squady gets all the reports of the squadron and knew none of these blokes were actually SIQ. He had hardly put the phone down and was on his way to their cabins on the second floor with room orderly in tow. Every room has two keys. One is with the cadet and the other with the room orderly. AK and Wallia had their rooms opposite each other on the second floor. Hearing the footsteps, Wallia stood on the chair and carefully peeped over the door through the seventh heaven (wire mesh). He could see the loyal orderly Hazari walking towards his cabin. Seeing him Wallia said, *"Abe tujhe bola tha Lunch break se pahle mat ana, koi Div O dekh lega."* (I had told you idiot not to come before the lunch break, some Div O will see you).

Soon the Squady walked in his field of view. Wallia fell to the floor like a thick lump of shit toppling the chair in the process. In the other room, AK Singh was shaken up from his deep slumber by the mighty crash. He heard the Squady asking the orderly to open the cabin door. Thinking fast, AK decided on the next course of action. Wallia's room was opened, and there stood Wallia in his underwear like a dressed chicken. "Where is AK Singh?" He demanded. "I have no idea sir," replied Wallia.

The orderly was told to open AK Singh's room. AK was nowhere to be seen. Wallia heaved a sigh of relief. The orderly peeped out to see behind the window as the Squady stood at the entrance of the cabin watching impatiently. Surely, he could not have jumped from the second floor; the orderly thought.There was no way of disappearing from that height. He then opened the cupboard and almost screamed in panic. In there was friend AK, rolled like a baby in a womb, hiding behind the hanging uniforms, holding his breath. He had his finger on his lips and with the other hand was signalling to the orderly to keep quiet. Unfortunately, the Squady had seen the orderly startle as he opened the cupboard and understood. He walked up to see AK Singh crawl out of the cupboard, sheepish and a bit embarrassed.

Both AK and Wallia stood outside in their undys (under wears). "Who locked you in the cabins?" asked the Squady looking at Wallia. "He sir," replied Wallia looking at AK Singh. "And you?" asked the Squady looking at AK Singh. "He sir," replied AK, looking at Wallia. "OK", said the Squady, not registering the mischief.

"Seven days restrictions, starting this evening," he announced and gave them ten minutes to join the class and report to Mr. Keshwani.

As they entered the class, Mr. Keshwani welcomed them with a smile. "Good to see you in the class. I did not mean to disturb your sleep but was missing my friends from my beloved Lima squadron. I hope the Squady did not punish you guys?" he asked.

"He did sir, seven days restrictions to each of us," replied AK Singh.

"How dare he punish my cadets. He is not smart enough to catch you without my help, so he has no right to punish you. I will ensure your punishment is cancelled." As he said so, the siren wailed signalling the end of the period. He walked out of the class straight to call up the Lima Squady and had the punishment order cancelled.

Mr. Keshwani was an icon who taught many generations. He was as much loved as he was respected and no one dared punish his cadets. Whenever NDA is talked about, he will always be 'Mentioned-in- Dispatches'.

Take Them Seriously at Your own Peril

Place- Science Block, Course- 48 NDA. A physics class was on for the fourth termers on the ground floor in the Science block. The Principal, a dignified quiet civilian, was taking a walk in the corridor visiting classes. Through the window he saw an unusual spectacle - a pair of shoes facing the roof. A cadet being taken to task? Wonderful, he must have thought. The civilian instructor had to be a tiger to be able to get a NDA cadet on his head in the class.

Cadet TN from Manipur was not feeling too bright. Possibly, the sixteen toasts that he had pushed down his throat-with numerous bowls of porridge and eggs were making their effects felt. More likely, the science class was too much for his limited talent. Not bothering to worry about the teacher, he rested his head on the desk and went to blissful sleep.

Soon the civilian instructor caught up with him and woke him up. "What happened, are you not well?" he asked. TN said he was having a severe headache which often bothered him, clearly a pill, as they say in NDA. The instructor suggested he should do yoga and in particular "Sheer shashan" (yoga exercise to stand on your head). Cadet Karnail, a burly non intellectual surd, who possibly never had a headache in his life, asked the instructor for how long should this be done.About fifteen- twenty minutes, replied the instructor.

"What are you saying sir, twenty minutes ! Not possible sir, no one in the world can do it for twenty minutes," said Karnail.

"What? You are saying no one can do it for twenty minutes in the world. I do it every day," said the instructor from Kerala proudly.

"Impossible sir, we can't believe it till you show us," said half the class in unison, pouncing like a pack of hyenas sensing the prey.

"OK, if you don't believe me, I will prove it to you," said the instructor, not realizing he had been tricked into a trap. He then stepped on the dais in front of the class, kneeled, cupped his hands on the floor, placed his head in his hands and slowly straightened his legs with his feet facing the roof. And the cadets, always up to mischief- keeping time ... "twenty... nineteen.... eighteen, well done sir, keep it up."

In walks the Principal, his expression simply indescribable- eyes dropping out of the socket, not quite believing what he saw- the teacher on his head and the class on their seats. He probably had thought it was a cadet under punishment. The teacher sheepishly unwound himself from his yoga posture and stood embarrassed. The Principal walked off without a word. As for the state of the teacher, the less said the better. You can imagine for yourself.

Bravo cadets, great show.

Catch me if You Can

Place -Sudan Block, Lecture hall number 99, Course - 48 NDA again. Another day, another set of cadets and teacher, another classroom but fun all the same. There are set dialogues for describing different things in NDA. When a senior bullshits a junior or asks him to run an errand, they will finish by saying, "Make it bloody quick and fast, OK."

A history instructor, another civilian, was teaching the class rather seriously- that no one was listening was another matter. Half way, he said, "I will now tell you about the Indo - China equation." Cadet Patra, another fun loving cadet , snapped from one end of the class "Make it bloody quick and fast, OK."

Bad luck, the teacher saw him. "You bloody idiot, I will teach you life," screamed the teacher and dashed towards Patra pointing his finger ominously. Patra, an athletic bloke, smartly took off over the desk before the teacher could get to him. And then started the Tom and Jerry chase - teacher chasing Patra in the aisles of the class- Patra keeping just out of reach with a 'catch me if you can' smile. And well, the class bucking up the teacher " Sir you are there, almost got him, buck up-buck up- just step out a bit and you will get the clot." The chase went on for a while as Patra smartly glided his way between the desks.

The way the episode concluded can only be described as 'poetic injustice'.

As the luck would have it, the same old Principal happened to be walking in the corridor. Hearing the commotion, he walked into the class. As quickly as it had started, there was deathly silence as everything came to a standstill abruptly. Rather confused at seeing the spectacle, he yelled "What's this happening Mr. XXX?"

Mr. XXX, shocked by the sudden appearance of the Principal, was dumb struck and then surprised everyone "Sir.. er.. er, sir Cadet Patra is a very good cadet."

There was very little for the Principal to do as he walked away shaking his head in disbelief. The class got back to Indo - China equation as the cadets laughed their guts out.

The Gods are on cadets' side more often than not and why not- they are the ones who give their lives for the country – without a blink.

Kuita Takes the Cake

The NDA trains cadets from a large number of countries. Majority of these cadets did not have the appetite for the kind of ragging the academy was famous for, neither were they subjected to ragging by Indian cadets. Of the foreign cadets, the Tongans were the most spirited and well built and often joined the 'Screw parade'. It was not uncommon to see my course mate Tony, a solid hundred kilo in weight walking down the road with three cycles round his neck, locked together with a chain. His course mates would give him their cycles to avoid 'Saturday screw parades' which was a torture with the cycle on the head. Tony would walk back to the squadron as if he was on a stroll.

It was in the Kilo Squadron that a third termer ran away because of ragging. Kilo squadron had always been as famous for ragging, as for its 'singing style' of making announcement in the squadron. It was quite uncommon for a cadet to run away at that late stage in the third term. So it was felt that there was indeed something serious, and the squadron commander, called 'Squady' in the academy, was asked to investigate. The Squady being ex NDA himself knew well enough that no Indian cadet is going to crib, to use the academy term. He decided to work on a Tongan cadet called Kuita, a third termer himself. Kuita was a sweet Tongan, loved and known all over the academy. Short and light for a Tongan, he was over eighty kilos with an impressive paunch, a roly-poly character with a shaven head, which he started sporting ever since his tipsy country mates shaved him off after a drinking binge. Kuita apparently impressed with his new found looks carried on that way for ever after. His shaven head and a flattish nose gave him a wonderful sweet look.

So the Squady paid a Sherlock Holmes visit to the squadron and started talking to Kuita. The conversation went like this:-

"How are you, Kuita?"

"Very well, *tank you,* sir." (Speaking in heavy Tongan accent).

"How is everything here? I hope you have no problems."

"No, none at all, sir."

"I hope nobody is troubling the cadets, ragging etc."

"No sir, life couldn't be better. No ragging at all in the squadron ever, sir."

Squady knew it was not quite true. Kilo squadron, he knew was notorious for ragging.

"Why did your course mate Joseph run away Kuita?" asked the Squady.

"I don't know sir," replied Kuita seriously.

"No Kuita, I am sure you would be knowing why Joseph ran away. After all, you are staying in the squadron and Joseph was your course mate."

"That is right sir, but I really don't know. No I don't know why Joseph ran away sir."

Squady pressed on, knowing Kuita was his best bet to get to the bottom of the case and he had to answer the commandant on the issue.

"Kuita, don't worry, nobody will be punished. I just want to know what the problem is so that we could sort that out."

Kuita then thought for a few seconds and said, "Sir, you really want me to tell you?"

Squady suddenly saw a ray of hope and expectantly said,

"Yes – yes, tell me Kuita."

"But I won't tell you," replied Kuita half singing with a mischievous smile and looking at the Squady from the corner of his eye.

"Kuita, don't worry about anything, tell me. I promise no one will be punished," said the Squady a bit disappointed.

"Sir, you really want me to tell you?"

"Yes -yes Kuita." The Squady was beginning to get restless and irritated but kept his cool.

"But I won't tell you," Kuita sang again with a bigger mischievous smile enjoying the fun.

"Oh ! Come on Kuita. I told you I won't harm anyone, tell me why did Cadet Joseph run away?" said the Squady truly exasperated.

"Sir, you really-really want to know why he ran away?"

"Yes Kuita," said Squady holding his breath.

"Sir, he ran away because…." and paused before continuing as the Squady gazed at Kuita's mouth.

"Yes Kuita, go on," the Squady egged him on.

"Sir, he ran away because – because – because - there are too many bed bugs."

Squady was crest fallen, seeing his prolonged, patient effort crash land into a total fiasco. Kuita lived up to the motto of the NDA, which is 'Cribbing is un-officer like'.

Well, as far as the bed bugs are concerned Kuita had a point. Joseph may not have run away because of bed bugs, but they were a menace of monumental proportion in the NDA. It was a test of wits between the cadets and the bed bugs. The bed bugs literally got after the cadets' blood. Cadets had rashes all over. Then came the brilliant ideas; the cadets put their stinking sweaty socks and vests in the kit bag and put it to sleep on the bed while they themselves slept peacefully on the floor. The trick worked to perfection, for by the time the bugs realized that they were fooled, the cadets were up and gone. Those days, every Sunday the cadets would take their mattresses out and give it a hot water treatment to kill the bugs. That did not work though and then the idea of a brilliant deception plan germinated.

Well, Kuita's story was not over yet. One year senior to him in the same squadron were his two country men, my course mates. One of them was a naval cadet as was Kuita. Suddenly Kuita applied for a change of service from navy to army. The foreign cadets normally did not do that, as in their case, their country finalised it before the cadets joined the academy in India.

The Squady asked him his reasons for doing so. He replied, "mine is a small country which has just started a navy with two boats. Cadet Tony, my country senior is one year senior to me. He is also a naval cadet. He will become the naval chief soon after commission and I will become the vice chief and remain vice chief for ever and never become chief, so I want to change my service and join the army." He indeed was a short man with a far sight.

Afzal and his Lanyard

As mentioned before, passing the drill square test in the academy is a great achievement in the life of a cadet. It is a difficult test to pass as well. How difficult it was, I realized only after I had gone through the whole experience. More about that later. Almost no one ever in the academy passed the drill test in the first term. The better lot passed in the second term. Majority managed in the third and the stragglers passed in the beginning of the fourth term. There was a general amnesty by the fourth term midterm break, because it was understood that those who had not passed so far, will never ever pass.

The academy has a lot of foreign cadets from all over the world. There were cadets also from Afghanistan. One of these was Cadet Afzal. Foreign cadets could not speak Hindi, some were not well versed with English either. Hence they were put through a language course in English & Hindi during the first six months. As a result of this, they lost a term and started their first term when their course mates went to the second. That isn't a good feeling at all.

Afzal with the passage of time found himself in the fourth (actually fifth) term, but had not passed his drill square test yet. Having got fed up of tying an arm band, he decided "to hell with the drill square, it has been one hell of a long time with the god damn arm band. I cannot be the lone fourth termer wearing an arm band." Since these issues are brought up in the periodic interviews of cadets, all officers were aware that friend Afzal was yet to pass the drill test. One day the Div O was surprised to find Afzal saluting rather than pulling his arms. He noticed that he was wearing a lanyard as well. Afzal was warned not to disobey the academy orders again. Afzal however was not impressed. Sooner than later, the same officer saw Afzal wearing the lanyard again. This time the officer ordered him to report during the breakfast break in what was called 'Bajri order'. As the name suggests, this was a rig in which the cadet wore an overall and carried a pack full of sand and stones on his back. Breakfast break was the most popular time to call cadets, as apart from the punishment, it hurt the cadet most as it screwed up your breakfast, the most sought after meal of the day after a gruelling outdoor training.

Afzal, however, had had enough and had his plans chalked out. He reported at the given time in overalls with a big pack full of stones on his back. The officer was surprised to see that he was also wearing a lanyard, which was obviously not part of the dress and never done. The officer, amused though, was annoyed at this insubordination.

Afzal was marched up to the Adjutant of the Academy, Major Harjit Singh of Brigade of The Guards. Having been marched in to the office, this time without lanyard, he was asked by the Adjutant.

"Afzal the charge against you is that on 14 March you were warned by your Div O for wearing a lanyard without passing the drill square test. In spite of that, you continued to wear the lanyard. When asked to report to the officer, you reported in the Bajri order with a lanyard on. This is a serious disobedience of orders for which you can be seriously punished. Do you have anything to say?"

"Yes sir," replied Afzal and continued, "Afzal is in the fourth term and his course mates are in the fifth. Afzal is the only one in his course who wears an arm band. Afzal does not know what the NDA has taught him. After your teaching me drill for five terms, Afzal has neither learned Indian drill, he has also forgotten Afghan drill. This way I will pass out of NDA without passing the drill test."

The Adjutant looked at Afzal. Afzal's childlike innocence struck an emotional chord with the Adjutant. After pondering for a while, he said with a smile, "Well, I see your point. Afzal could not pass the drill test in the drill square but he passes it here. You are hereby declared passed in the drill test which never took place. But, for the disobedience of orders, you are awarded fourteen days restrictions." Miracles do happen in NDA. Afzal could not believe his ears. He had never even dreamt that it could end this way. "You may leave now," ordered the Adjutant.

And then, putting his hands in his pocket, Afzal pulled out a lanyard, wore it on his shoulders, smartly saluted the Adjutant, thanked him and marched out with a wide grin. Afzal was seen happier than he had ever been. And guess what, the first Sunday, he proudly applied for liberty and was seen waiting for the bus to Pune, his first trip out in two and a half years.

I Should have Known Better

As I said earlier, passing the drill test in NDA was no mean task. Let me take you through a drill test, just to tell you what I mean. Towards the end of our first term, we were given an opportunity to appear in the drill square test. I thought I was reasonably good at drill, as did many others. So I decided to make a serious attempt to pass the drill square and get at least one worry off my head, the others being the hand spring and upper circle, not to mention the ten metre jump in the pool, quite literally waterloo for some.

I put in best efforts to prepare for it, spent a lot of money to get my turnout perfect, or so I thought. My orderly Banwari spent seven days shining drill boots and they shone like a mirror. Brand new shoe laces to ensure there were no turns and twists, ensuring there were thirteen new hobnails and one spare in the pocket; just in case one fell off at the nick of time.

Everything was simply brand new being worn for the first time. No chances taken. The cotton drill shorts and shirts were so well starched that they would stand erect, when dropped, as if made of hard board. The shirt would badly bruise your arm pits when you swung your arms.

Finally the big day had arrived. Under the supervision of the fourth term corporal, we marched to the drill square with the shorts and shirts in hands and the boots round our necks. Sure, I was ready. The uniform was taken to be worn at the drill square to prevent it from getting crumpled. You could take no chances with the drill instructors. The corporal barked out the order to fall in. He carried out his initial turn out inspection and said it was hopeless and none stood a chance. What a prophecy! Only time would tell. Soon we were lined up in the drill square for the drill test, twelve squadrons each of twenty five cadets.

Turn out inspection was the first part of the test. Drill test would follow later, that is, if you cleared the first hurdle. There was no officer to oversee the test. The drill Subedar Major was the first and the last word. He was the emperor of all he surveyed on the drill square and his judgement was final. The drill Subedar Major started his turn out inspection. He did not take long to run through the course, which he did like a red hot knife through butter. At the end of his summary disposal we were thirty cadets left out of two hundred and fifty. Most were not disappointed; they did not expect to get far any way. The thirty of us, who had survived the genocide, were lined up for a detailed look. The Subedar Major started the ritual from one side. In no time he was through with us. First in the line was dismissed for not shaving his moustache. He did not have one. The second fell from grace for smiling on parade. He had a typical ever smiling look on his face which had got him into trouble earlier as well. The next was dismissed for having one of his thirteen hob nails missing. He produced one from his pocket saying that it had just come off seconds ago. It was an old excuse that got him nowhere. Many cadets carried a spare hobnail during the drill periods as a missing hobnail earned them extra drill parades. One cadet was packed up for looking down when the Subedar Major told him "Look, there is a twist in your shoe lace"; next was dispatched with disdain when he did not look down when the same fault was pointed out which did not exist.

At the end of the exercise, we were three cadets left. It must go to the credit of the Subedar Major that for the twenty seven dismissed, no excuse was repeated, and most were not true yet indisputable. It was quite clear to all of us that the drill test was a farce, an exercise in futility. We realized the end was not far, yet, a man lives on hopes and thought we three may just be the exception.

Finally he lined up the last of the survivors. I was the last in the line. The Subedar Major shoved his little finger under the belt of the first cadet. He was dismissed for a loose belt. The second, hanged for a faded button. He then stood in front of me, the last and the lone man standing in that vast magnificent drill square, watched by a dozen drill instructors and said,

"You have come a long way cadet. I am sure you would have worked very hard to get this far. We got to wait and see what is in store for you now." He surveyed me from head to toe. Stopped to look at the tightness of the belt. I seemed to have passed that test. He took his gaze further down and surveyed. All the while, I was careful, not to even blink. So far so good.

Nevertheless, I had a sinking feeling that if two hundred and forty nine cadets had been given marching orders, there was little chance that I would be the exception. But I am an optimist, and yes, a man lives on hope. I was sure he was working out a brilliant idea to nail me. He circled around me a few times like a tiger on the prowl and stood behind me for what seemed to be a long time. Harnessing his thirty years experience, he critically examined me. Suddenly he broke the silence and barked, *"Cadet baen paer upar - fail, line tore,"* he shouted with a sadistic pleasure. (Cadet lift your left leg- fail, break off). The coup- de- main was delivered; ***I had not polished the sole of my boots.***

So that was the end of our honey moon with the drill square test. In what was supposed to be a drill test, none out of the two hundred and fifty cadets ever got to the drill stage. Despite the best turn out on earth, all failed to get past the drill Subedar Major. We had actually wasted our time and money, but we had no malice towards him. I am certain, he was only obeying Adjutant's orders.

The Battle of the Basketball Court

The battle of Waterloo was won on the playing fields of Eton. Sports hold special importance in the military for they build morale and camaraderie. In the NDA too sports have a special place and a number of inter squadron sports competitions are held every term.

Inter squadron competitions are fought with a do or die attitude. All squadrons put in everything to win. At the end of the term the squadron standing over all first is awarded the championship banner. The perks - it is the only squadron that wears lanyard on the right shoulder, while all others wear it on the left. The other perk - standing first in the line up of the squadrons in the drill square. Standing at one end of the drill square, like the last squadron at the other end, they do not have to bang their feet, (The NDA drill square is so long that stamping of feet by the squadrons on the extreme ends shatters the sound of drill movements). The squadrons in the run for the championship banner work out all sorts of strategies and tricks in the bag.

It was around my fifth term in 1977 that a revolution was introduced in the inter squadron competitions. Then on, the squadrons were required to field six teams each, called six strings, for every sport competition, as against two earlier. In a squadron of one hundred and twenty, producing six teams was no joke. It meant that every bloke was required to play whether or not he knew the ABC of the game. Some games like football did not really matter; raising six teams for the more technical games like basketball was a problem indeed.

It therefore turned out, that in each squadron the sixth string team invariably consisted of real wrestlers, who had no clue on earth of the game. Every competition in the sixth string turned out into a no holds barred wrestling match, where, the strategy was man or the ball. The players invariably chose the man, for he was easier to grab than the eluding ball. The casualty rate in sixth string matches were always high.

It once so happened that the result of inter squadron Basketball championship finally hinged on a sixth string match between Echo and India squadrons. The squadron that won this match would have won the overall championship as well. If it ended in a draw, then some other squadron would emerge the overall champions. This then became the match of the season, the most important match between two teams that knew nothing about the game.

The match took place drawing the largest crowd of spectators as half of the academy turned out to see the game. The score in that Basketball match till the breather was zero all. The score till the last three minutes was one point all, scored from dozens of free throws awarded to both teams. Casualty state in a game of no physical contact were six teeth, two swollen eyes, three split heads and many walking wounded who continued to battle it out in the court 'beyond the call of duty'. With rapid casualty rate, both team's reserve on the bench was fast depleting, threatening to run out of players.

As the clock ticked away, the josh and the violence grew. In desperation to score a basket, the ball was thrown at all angles and range that would have amused Michael Jordan. Most balls missed the board altogether, some hit the board like a rocket and ricocheted even faster threatening the safety of the spectators. An odd one hit the ring threatening to take it apart, drawing wild applause from the gallery, for that was an achievement. The cheering from the competing squadrons was deafening. In the remaining last three minutes, neither team failed to trouble the scorer. Being a championship decider, the game went into extra time as the tension mounted.

During the change over break, expert comments to both the teams came unsolicited, full throated. Some threatened the team of dire consequences if they lost. That was the NDA way of motivation, except that the players understood nothing and cared even less. They were just raring to go, not for the ball, but for the man, for the squadron's sake. When all this was going on, Echo squadron, the last term's champion squadron, quickly drafted masseurs (remaining first termers) to massage the players for the final assault on the court. It looked more like a wrestling arena and preparation for the boxing bout.

Finally, both the teams adopted the strategy *"Forget the ball stupid. Go for the man; ball cannot go into the ring by itself."* The logic was simple, it was the man behind the ball that mattered. If you can injure all the opponent players out of the game, you would have an empty court to score the winning basket.

In this clash of titans the referee became a non entity as the law was taken into their hands by the players. If the referee blew a whistle for a foul, the team that grabbed the ball first took off with the ball.

An ambulance with a Nursing Assistant (NA) was stationed to provide medical cover. The NA soon realized he had more than he could handle on his own and asked for reinforcements.

A medical officer with two NAs was rushed to his aid. Serious casualties from the sports battlefield were ferried to the military hospital. The 'walking wounded' were managed at the medical aid post established in situ and drafted back in action.

By now, both teams were down to three active players on the court each, the others raring to join the battle after first aid, for which there was a queue. The doctor, a Major, had been in battle before, but this was too fast, too furious and too bloody. The referee announced last sixty seconds. All hell seemed to have been let loose. The players went crazy, charging around aimlessly, going for anyone who happened to be anywhere near the ball. Suddenly, the star player of Echo Squadron, recovering from a temporary black out after a head on collision, roared into the court like a warrior with a bandaged eye and jumped into the fray with the zeal of a Samurai determined to decimate all enemy.

Then, in the last few seconds of the game, someone from the gallery shouted, *"Last five seconds to go, just throw the ball, you bloody monkey."* The last Samurai, facing backwards threw the ball from short of the centre line. The ball soared into the sky, apparently heading for space. All seemed to be over as a draw would have given the championship to a third squadron. As it began its descent, the ball seemed to be heading for the ring. Suddenly the deafening roar subsided falling into an uneasy silence as everyone stood quiet watching the ball. It seemed to be coming down in slow motion, as if the Gods wanted the game to go on a bit longer. The ball came down crashing, landing at the base of the ring and shooting right up into the sky. The ball bounced a number of times on the ring perfectly straight up and down, as if by God's will, as the cadets watched the spectacle wide eyed. The championship match had taken its toll in blood and broken bones, not to mention buckets of sweat. Finally, the ball swung into the ring, circling for what seemed to be an eternity, threatening to roll out any time. As if to mark the grand finale, the ball swung out of the ring and came to a standstill on its neck. The academy watched in utter disbelief and in complete silence. And as the players unsuccessfully jumped up and down to reach the ball, perched beyond their reach, the last Samurai was seen climbing on the shoulders of his team mate to reach the ball. While this spectacle was unfolding, another team mate was trying to shake the Basketball pole hoping the ball will roll into the ring. Seeing this, a Texan from the opposition grabbed him by the waist and in one big pull dragged him away, both hitting the dirt in the process. Before the tableau could be completed, the ball, as if by God's hand rolled into the ring. The academy went up into a deafening roar. Echo squadron invaded the field and lifted the Samurai on their shoulders, the wounded architect of their victory, who saw them lift the championship trophy. India squadron watched in stunned silence, unable to believe the way they lost. Finally Echo squadron won by one basket, final score reading 3-1. The scum of the earth, the first termers of Echo Squadron had become instant heroes. For the first termers of the losing India Squadron, who had sweated and bled no less, the misery had just begun. They were asked to report in the squadron bathroom in next ten minutes. The rest is history. Some match it was.

In a match of no body touch, the casualty state sounded as if from a platoon quick attack on Kargil heights. The doctor himself barely survived nervous breakdown. The story of this cracker of a game was remembered for a long time. While the final score of the Basketball game, 3-1, sounded more like a soccer score, the casualty figures looked closer to a Basketball score, nine all, with many recycled. It is this indomitable spirit of the NDA that wins us the impossible wars. Bravo NDA. It surely was a game sculpted by God himself-the mother of all games.

EVERY DROP MAKES AN OCEAN

The NDA cross country is an event in itself. Over one thousand five hundred cadets in multi- coloured squadron vests line up in the unending glider dome. Last minute motivation talks by the Squadron Cadet Captains (SCC) and then the whole academy taking off as if there is no tomorrow is a sight to behold. Add to that scores of officers smartly mounted on well bred horses cantering along. How strongly the cadet feels he was on the horseback, but there is no getting away.

The cross country route was pretty taxing with the 'Rape Hill' adding salt to the wound. The squadrons worked very hard to do well, this invariably being the first competition of the term. I was the SCC in Lima Squadron, called the Lions. We had worked bloody hard-running before muster and after dinner. On the day of the competition the squadron fell in before moving to the glider dome. I pepped up the squadron, saying that the time had come to put in their best and deliver. I said that everyone had to cross at least one cadet in the last hundred metres. One second termer tried to put me wise saying, how does that matter sir, it's enclosure system so you get the same points whether you come first or last in the enclosure. His logic was irrefutable. I told them this story why every bit of extra contribution was important.

Quote "There was a surd, a Jat and a four feet nothing highlander who went trekking in Africa. They lost their way and strayed into a cannibal tribe territory and soon were caught. They were brought to the tribal Chief who was up in rage. The Chief ordered a bon fire with the whole village in festive mood performing tribal dances. He ordered a feast with one man being barbecued at a time. The three musketeers were trembling with fear, certain their time had come. They fell on the Chief's feet and begged for mercy, pleading they made an unintentional mistake and would never be seen around ever in Africa. The Chief decided to give them a chance under one condition. He said their 'tools' would be measured the next day at dawn. If they measured twenty inches or more then he would spare them or else they would be eaten alive.

Having no option, they agreed. The night passed- somehow. Next morning the whole village had collected at dawn. Three of them were stripped naked and taken in front of the Chief. The Chief ordered the measurements to be taken. The surd measured nine inches. The Jat measured ten. Pretty close, they were certain they would be through. The highlander stepped forward and measured one inch. Twenty inches, declared the chief's man. Three of them exchanged glances with great relief, keeping everything crossed. The Chief, as promised, released them under one diktat. They had one hour to disappear from his territory. He would send his men after them and if they were caught, there would be no mercy. They ran like hell for two hours, and when they were certain that they were out of his territory, sat down to catch their breath.

The Jat claimed all the credit for saving their lives. The surd said his contribution was no less and claimed equal credit. The highlander listened silently. He then said, "You are claiming all the credit for saving our lives. Mine is the most important contribution in saving your lives." Both the companions laughed and asked him how his contribution was the most important? The highlander replied, "*Agar aine mauke par mera kh*ra, nahi hota na toh tum shaare mare jaate.*" (If my d**k had not risen to the occasion at the critical time, you would be all dead).

"So Lions as you can see, every inch of contribution is crucial," I said. The squadron in high spirits responded by a loud yah-yah. The Lions were convinced by the story, and when the cross country competition was over, the squadron had finished eleventh out of twelve squadrons.

So much for the moral lecture. As for the result, can't really blame the squadron. It was probably the only time in history of the NDA that there was a re - run in the academy. In that maddening commotion of the last hundred metres rush, the second and third enclosure types had got mixed, and many others, wanting to get their squadron extra points had gate-crashed into second enclosure leading to another run over the Rape hill.

Interestingly, as the SCC of Lion squadron I would say "NDA is a jungle and Lions rule the jungle- You are the King of this Jungle, so bash on regardless." Both these stories, I am told survived more than two decades after we left. Traditions die hard, as they say, however silly they may be.

The Unforgettable General Rider Sidhu

The cadets come to NDA from a variety of backgrounds, each with his own typical characteristics. There is the academician variety, always studying hard. Then there are others who would not touch a book with a barge pole. There are others who live up every day, without a worry in the world. Together they make a great collection of wonderful characters and add spice to life.

A course mate, affectionately called 'Rider Sidhu' was a lovable guy and a good rider, hence the sobriquet. Sent to Echo Squadron on joining the academy, Sidhu soon realized he had to create his nuisance value if he had to survive in the academy. There was one particular fourth termer bent upon making Sidhu's life miserable. Sidhu realized he had to launch a counter attack for the long term good. Sidhu strained his brains more than ever in his life and had his plans ready.

The next time the sadist fourth termer came to his cabin, Sidhu yelled *"You rascal, I will electrocute you and send you to hell,"* and in a well rehearsed move, yanked out the table lamp with a broken bulb and lunged at the fourth termer. The fourth termer ran for his dear life with Sidhu on the chase, lamp in hand. The cat and mouse chase continued as the senior, yelling for help, ran from one floor to another. Sidhu, allowing him to maintain the gap kept shouting, "Bastard, I will electrocute you." Hearing desperate cries, his course mates ran out of the cabin and came to his help. Sidhu was taken under control and paid for it. Thereafter he earned himself the sobriquet of *'Pagal Sidhu'* (mad Sidhu). Sidhu's reputation spread like wild fire and that was the end of all his woes in NDA forever. No senior ever again took the chance of getting electrocuted by a crazy surd.

With all his problems over, Sidhu was happiness personified. With no worries in life, he enjoyed every single moment in the academy. Already a 'General' (relegated twice) by the fourth term, he had a tremendous attitude and zest for life. Together both he and his brother could have easily qualified for Guinness book of world records for becoming 'Generals' in the shortest time. Sidhu was in trouble in Echo squadron with the Squady. Already a General, he wanted to come to Lima squadron to have a fair chance of passing out. My request to our wonderful Squady, Wing Commander Sayyad Ehsan Ali brought him to Lima Squadron.

Sidhu spent most of his time in whites, which is the dress for the sick cadets who are excused outdoor training on medical grounds. He bunked as many classes as possible. If he had a choice, he would have liked to pass out without attending any at all. He did not attend many, and when he did so, he was working out how not to attend the next, least bothered what was happening in the class.

Very often he would be seen playing cycle polo when he had his right hand index finger in a plaster. Everyone admired his determination and love for the game. Sidhu had mastered the art of leaving the class ten minutes before time during the breakfast and lunch break. He would simply get up, walk up to the civilian instructor, whisper something in his ears and walk out of the class. All this to beat the sergeants who took posts along the road. Others who tried the same stunt did not succeed.

Anyone asking his secret, he would say, *"Abhee tum bachhe ho, yaar"* (you are still kids, dude). We were good friends and when I asked him the same question, he shared his secret. He said look at my hand. There was nothing wrong with his hand. With a typically wide Sidhu grin, he put his hand in his pocket and when he took it out, his two main fingers were in plaster. Then, putting his hand in the other pocket, he took out a bandage sling, put it round his neck and hung his hand on it. Amazing brilliance, isn't it? And putting on an expression of deep misery, said, "Don't I look like a walking wounded from the battlefield? This is my 'Ram Baan', one solution to fit all problems. With civilian instructors, the plaster is good enough, the sling is needed to fool the sergeants," he said. And then, whistling, pulled out the cast and put it back in his pocket to become fighting fit again. Sidhu was great in his own way.

During one out door military tactical class on indication of landmarks, rider Sidhu was taken by surprise by a question from the Div O. He was asked, "Sidhu, did you see the last landmark I indicated? Check back."

Without even looking up, Sidhu replied "Sir, reference the last landmark indicated by you, six o' clock own position". The Div O conducting the class was from Corps of EME, who deal more with equipment management and rarely with tactics, said, "Good". That was quick thinking coated with super confidence. Sidhu got away, yet again, as he often did.

Good at heart, Sidhu had his fun with juniors but never harassed them, as some others did. Amongst his many original accomplishments was a 'musical orchestra' that he had created. It consisted of eight first termers whom he had trained as musical instruments. Each of them was trained to produce different types of musical sounds varying from tabla, keyboard, drums, guitars etc. He had trained them to play in two modes, the auto mode and the manual mode. To activate the orchestra, he would poke his finger in the stomach of the first in the line, who would then twist the ear of the one next to him and so on. The last one would produce a ping to indicate that the orchestra was all tuned and ready.

The orchestra would be lined up by Sidhu in a musical logic that only he understood. Every Saturday evening he would organize a musical evening by his band on the top floor flank of Lima squadron. Mostly, the captive audience consisted of first and second termers, even though he had a good course fan club too. He would produce different tunes by poking his finger in the stomach of the human instruments in a well rehearsed pattern, and incredibly, produced music pretty close to the original. His captive audience, as instructed by Sidhu, clapped in deep appreciation of the performance after every tune.

On the auto mode, activated by twisting the ear of the drummer, the orchestra would play 'Colonel Bogey' and 'Auld Lang Syne', the popular tunes on the drill square.

Bravo Sidhu. God needs good people too. He passed away young in a car accident and we miss him nostalgically.

No Laughing Matter

Pyush Shandillya was a fun loving character, always up to some mischief. In the fourth term we were all in class 54/4. He used to carry a laughing toy wherever he went. It was activated by pulling a string. It had a very sharp funny laughter- "Yeh – Yeh – Yeh ….." variety. The longer the string was pulled, the longer the monkey faced toy laughed.

He would play the trick with the civilian academic instructors whenever a joke was cracked. Usually the noise in the class lasted longer than the toy's funny laugh. Once it so happened that he pulled the string rather long. The laughter of the class subsided but the toy carried on laughing creating a real funny situation.

The civilian instructor, I still remember, was thoroughly confused and wondered who was laughing. He looked at the class, but the cadets sat mouth shut and expressionless. Of them all, Shandillya was the most serious. The instructor then walked to the window and peeped out looking for the source of laughter, there was no one. He went around the class looking for a clue but found none.

Shandillya was hardly the one to back off, and did it very often, always getting away. I know he had a backup plan. If there was a likelihood of getting caught, he would have quietly slipped the toy in the neighbour's satchel, that wonderful bloke called Ganguly, known as Gangs in 54 NDA course.

Well Done Mate.... Actually Sick?

When I left NDA, it had twelve squadrons named Alpha to Lima. Later three more were added making it fifteen. There were cadets who spent a lot of their time in the Military Hospital and it was termed as the sixteenth squadron. Many sought respite from the daily rigours in the sixteenth squadron.

Life in the first two terms in particular was pretty tough. Added to regular double outdoors was the tough regimen in the squadron. It was quite common for cadets to report sick to avoid hectic schedule. They would come up with funny excuses. Initially, the newly posted doctors in the NDA hospital followed the maxim that any one reported sick was actually sick. Soon they realized that at the NDA, it was not quite the case, often it was the opposite. There were professional sick reporters, who spent most of their time in whites (the dress of the sick cadets). They had the knack of bluffing the doctors, convincingly feigning sickness.

One cadet wanting to skip a double outdoor reported sick with an excuse of running stomach. "How often do you go to the toilet?" asked the doctor, who by now had become smart. "Almost every hour," replied the cadet clutching his stomach as if in great pain.

"Alright," said the doctor, "Sit in the waiting room. I will like to see your stool. When you feel like going for a crap, let me know, we will provide you a pan."

Holy crap man, I had it, the cadet thought. He waited in the room wondering what to do next. After sometime, he approached the doctor and said he was ready. A pan was provided. The cadet produced a convincing sample. The doctor gave him 'Attend C' (rest in the room) for forty eight hours. The truth is, he never crapped. The secret, he had procured shit from somewhere, mixed it with water and produced a convincing sample and earned two days rest. Well deserved I guess. On another occasion a cadet reported sick, telling the doctor that he could not move his thumb left–right. When the doctor caught him, saying that he was in fact moving his thumb, the cadet thought fast and said, it was the other thumb that he was talking about.

Sometimes, when it became difficult for the doctors to decide who actually was sick, and who was not, some doctors would adopt the real infantry way. They would make all cadets on sick report run around a tree. The first few were given Attend C, the next few-Attend B (excused outdoor training). Those actually sick could hardly be expected to come first in a run. They sometimes landed up with punishments for feigning sickness. Long live NDA

A Blind Man Leading Another Man...

During the course of the training, cadets attend training camps in the second, fourth and the sixth term. While the second term camp- 'Green Horn' is meant to teach basics of camping – tent pitching, a bit of map reading etc, the fourth term camp-'Camp Rover' is a bit more serious and professional. The cadets love the camps, an opportunity to get away from the sadist seniors. The approach of cadets in the camps is best described as 'misguided missiles', all josh and no hosh.

Story of academy camps cannot be complete without the mention of 'Tipsy pudding'. Every camp terminated with a camp fire, with the cadets putting up some kind of entertainment show. The icing on the cake was the famous tipsy pudding. The cadets hogged till they were sick, and above all, it was packed and smuggled for the cadets back in the squadron. An academy folklore goes that the cadets have even swam across the Khadakwasla Lake with the famous tipsy pudding; some guts! An ex NDA thinks that he is the last word on tipsy pudding, just because he has eaten mess tin loads with his dirty hands in the academy. As a cadet he has the foggiest idea of what goes into making it, but his association with it is emotional that lasts forever. Serve him tipsy any time of life and he will find it hard to keep off from narrating the tipsy episodes from NDA, and how great it tasted, lamenting, he has never again eaten such wonderful tipsy. The worst victim of this emotional association, later in his life, is his unfortunate wife. She can never satisfy this bloke with tipsy that she makes, whatever she may try.

In the fourth term at the academy, there is a training camp- Camp Rover, aimed at putting the cadets through basics of military training. The camp is an inter squadron competition, with a variety of events such as tent pitching, preparation of defences, camouflage and concealment etc being part of it. The last activity of the camp is what is called the 'josh run', akin to endurance run. It is a twenty kilometre point to point navigation exercise back to the academy, and carries maximum points towards the competition. The squadron finishing it fastest, touching all the check points also gets a separate trophy and a lot of sweets. All the squadrons put in their best and run like hell to win this event.

During our camp, the competition for the championship was between Lima and Juliet squadrons. Squadrons were released along the same route at an interval of fifteen minutes. They were given the six figure coordinates of the first check point on a piece of paper at the starting point. The details of the second check point would be given at the first check point and so on. Juliet squadron preceded us and took off from the starting point running like horses. We noticed that none of them even bothered to open the map, let alone chalk out the bearing and distance to the check point. They blindly took off in the same direction as the squadron before them. My squadron followed suit after fifteen minutes. Our navigation team quickly worked out the bearing and distance and led the squadron on the double. Luckily for Juliet squadron, their blind faith in preceding squadron worked well. They reached the prominent first check point without using compass or the map, simply following the general direction of the preceding squadron.

One of the check points was a small trig height (a small hill). That point was proving to be difficult to locate as there were many such in the area. The old edition maps, our poor map reading, inaccurate compasses and the sweltering heat made matters worse. As we steadily moved forward guided by the compass, we saw Juliet squadron in full josh speeding way off the course. Caught in doubt ourselves, we stopped to re- check our bearing from the map. Two different sets of calculations confirmed we were on the right lines and we continued on our way, happy that our closest rivals were on a wild goose chase.

We reported at the correct check point, took the coordinates of the next and soon we were on our way. Our rival Juliet squadron was nowhere in sight. Thoroughly drained, we finished the endurance run and waited patiently for other squadrons to close in. We had won all there was to win, the josh run and the camp trophy. Finally, rather late, Juliet squadron trickled in one by one, the last squadron in.

The fate of Juliet squadron interested us. Later we learnt that they had ridden their luck for too long. They blindly followed the preceding squadron and rarely opened their map and compass. When the other squadron got lost on the tricky check point, Juliet squadron tried to figure out their way ahead, never easy when you are not even sure of own position. They knocked around for sometime without locating the check point. Cadet Rehal, who has been 'mentioned- in- dispatches' earlier, a sincere enthusiastic bloke was part of this group. He shouted pointing at a distant hillock,"Look there, you can see the Div O's cap on the skyline." "Yes, there is the Div O on the check point, come on guys on the double," shouted others and bolted off in that direction with full josh without looking at the map. Panting like horses and sweating like pigs in the hot summer, they scrambled up the hillock on all fours, their packs and water bottles dangling all over the place. When they got on the top, barely thirty yards from what had appeared to be the Div O in his field cap, a big *'langoor'* (white monkey) was surprised by the unexpected guests. Angered by intrusion into his privacy, he growled and bounded off with his companion. There was no sign of the Div O. From a distance, the monkey's head on the skyline looked like soldier's head with a field cap. They had actually taken off in the opposite direction, and climbing up the hill, were done and gave up the chase for the trophy. Cadet Rehal was the convenient scapegoat, even though everyone had agreed with him. The lapse possibly cost them the championship. And poor Rehal, looking for a hole to bury himself, would have to bear the brunt for the rest of the

term. Lesson learnt: look before you climb a hill and believe half of what you see. Further, men can be up to monkey tricks, monkeys cannot be men.

Camp Torna... you Son of a...

- *Div O – Divisional Officer of the rank of Captain responsible for the course.*

The camps in the academy, as I said, are a source of fun and frolic. We were in the final term in which you have a training camp by the name 'Camp Torna'. This camp teaches section level tactics including patrolling, laying ambushes, taking defences, siting support weapons etc. Having reached the camp site in the peak of summers after a gruelling walk, orders were issued and we were required to dig defences. Cadet Gerard Yong from Singapore was nominated the Section Commander for that leg of the camp.

The Div O, Captain Nagal, a tall soldierly officer was business like. His instructions were crystal clear. The trenches will be six feet deep before sunrise, no light or torches will be used and no one will sleep, even for a moment. He said he would make a surprise visit any time and anyone caught sleeping will be hanged by his family jewels. He added, he would ask everyone to stand in the trench in the morning. Any one's nut seen above the trench would be buried in the same trench without military honours. The Section Commander Gerard Yong was responsible to ensure that the section was fully ready by morning. After having had our meager dinner of over cooked, teeth breaking *'shakarparas'* and near boiling water from the Second World War design water bottles, we got down to digging.

The pick axes bouncing off the rocky ground sent sparks flying as the cadets cursed everything on earth. In the pitch dark night, the future looked darker. Soon I realized that it was a waste of effort trying to dig in this rocky ground on the hills of Mahabaleshwar. I told my buddy A V Kumar, an amiable 'tambi' from Karakootam, that there was no point wasting time. We won't get very far on this ground. I suggested we go to sleep and early morning find some softer ground and be ready for the show. To sleep is an offer, a good NDA cadet can never resist. He agreed even before I had finished. Making good use of the ground we found our way to the rest area, and hit the sack. The rest of the disgusted souls continued sledging and cursing away. The sound of pick axes banging away on the rocks and flying sparks mingling with the curses was like a good 'Son et Lumiere' show on the history of trenches. After a hard day's work that left us drained, they were a perfect lullaby, and before we realized we were fast asleep and snoring with gay abandon.

Soon, the Section Commander, having found us missing located us. He told us to get to work. We convinced him that in the morning he would find our trench as deep as the deepest in the section. This of course was based on the assumption, that, given the state of ground, no one would get very far.

Gerard's sight was not particularly an encouraging one. He already had blisters in both hands and had tied handkerchiefs. His pale skin was pinkish red with heat and fatigue. He gave an impression that the worst thing he had done in life was to join the NDA. Saddled with the responsibility of the Section Commander, he bravely trudged along banging away.

We must have been sleeping for about three hours. NDA's survival instinct told us it was time for the Div O to come and time we got back to digging. When we got back to our trench, the whole section was still digging, barely able to lift pick axes. With no energy left to speak, the cursing had disappeared. We were not surprised to see that after three hours of digging, the trenches were less than ankle deep. They had hardly made any progress, and had little to show for their efforts, save more blisters, sweat and dead with fatigue. Ten minutes on, everyone was dead tired and decided to take a break. They all disappeared in the darkness and soon fell into blissful slumber.

Not much later, we saw a dim, issue type torch moving in our direction that hardly worked. Possibly Capt Nagal, we thought, but did not care as we wondered what to do with this whole business of digging which had no future. Suddenly, in the pitch dark night, we heard our Div O's husky voice.

"Who are these jokers digging?"

"Sir, it's me and him," I replied.

"Who me and him, you clot."

"Cadets Sengar and AV Kumar, sir," I replied.

"Where are the rest of the clowns?"

"They were digging till sometime ago, just taking a break sir," I replied.

"Don't try to teach your pop and give me that crap. I know you guys too well. I left NDA when you guys were still in undys. Call the Section Commander and tell him to get everyone here in two minutes flat."

AV Kumar went looking for Gerard who was snoring away, sleeping like a log behind a bush. It took herculean effort to shake him awake. The moment he learnt that the Div O was here, he sprang up like a rifle magazine spring set free. Half groggy, stumbling and falling in pitch dark, Gerard managed to find the whole lot with great difficulty, for the cadets had made good use of ground and cover. Slowly every one strolled in rubbing their eyes and adjusting their dungarees.

They fell in hurriedly and the Section Commander gave all present report.

The Div O then gave his piece of mind with choice expletives. That was not new to us after three long years in the academy. "You clots, I had made it very clear that you were to dig the whole night and no one was to sleep. When I walk in, I only find two sincere blokes, Sengar and AV Kumar seriously digging and the rest of you bloody sleeping. You had the audacity to sleep despite my clear instructions to the contrary. You deserve a kick on your arse and you will get one right here. Sengar and AV Kumar stand aside. The rest of you, turn around and get rolling down the hill," finished the Div O. The Section Commander tried to plead that they were actually digging the whole time and had barely been resting less than ten minutes when he came. The Div O won't take that. He cannot be blamed. It was always difficult to tell whether a NDA cadet was fibbing or was being truthful, he looked so convincing all the time.

For the next two hours, the course kept rolling up and down the hills of Mahabaleshwar, while the two of us stood aside, rewarded for serious digging!

Then, the Div O, possibly feeling sleepy himself, got us together and said, "You will now dig till sunrise without a break and I mean without a break. Sengar and AV, I am happy with you, both of you can take one hour rest in between whenever you choose to," and then he departed.

Everyone cursed us while we laughed our guts out. Well, our one hour rest started the moment the Div O left and continued till sunrise, the longest hour. The rest got down to digging again, cursing everything on earth including the two of us.

Well, the next day, to our good luck, the Div O never went around the section post to see the state of trenches, as he had threatened the day before. If he had done so, he would have had to dig a trench himself to bury the two of us, since our trench was not even deep enough to bury a rat. Napoleon had once said: *"I do not want competent Generals, I want lucky Generals."* I guess, he would have been delighted to have the two of us leading his cavalry charges.

Later, having passed out from the NDA, we were in the

Indian Military Academy, Dehradun. Capt Nagal, the popular Div O that he was in Lima squadron, came to look us up. It was there that I told him of the story of Camp Torna. He enjoyed the episode, smiled and said, *"You Bastard."*

Three decades later, Capt Nagal, as expected had gone on to become a three star General. In 2007 I was posted at Ambala. After a war game in Ambala attended by the formation commanders of Western Command, they were invited to a break up dinner. General Nagal was there too. Surrounded by many other Generals, I told him I was compiling a book on military humour. One story featured him at NDA during Camp Torna, and wanted his permission to quote him by name. He said he did not remember the incident and wanted me to narrate it. I did and he had a hearty laugh. He then said, "Yah, sure you can mention my name. The only thing, whole army will say, look, such dopes can also become Corps Commanders. Any way, it will be a morale booster for many, so go ahead." As always, General Nagal had a great sense of humour.

The Academy Lingo

English was the national language of NDA. Speaking to a senior in Hindi was considered being 'Cheeky' and prohibited. Speaking even to a course mate in Hindi in hearing distance of a senior was asking for trouble. The public school types, speaking good English were snubbed and battered into speaking like officers. Many, particularly from Texas (Haryana) exploited the opportunity to use the captive audience of first termers to practice their oratory skills and improve their English. Kunjians, as the cadets from Sainik School Kunjpura were known, were conspicuous by their knowledge or lack of 'Foren language'.

A fifth term Sergeant Rathee was often seen punishing the cadets ordering them around, "You *bashtads*, go round that wooden tree and electric lamp posts". If he was not satisfied with the response he would tell them. "You *bashtads,* see me in bajri order today tonight, tomorrow tonight and day after tonight. Don't forget or you will remember your pope, ok?"

Hindi was rarely heard or spoken, except on the drill square or when attending Hindi classes. The English culture caught on with the drill instructors as well. They would sometime submit their report in English. One report leading to the award of punishment, complaining about a cadet slouching under a tree with hands in the pocket read, *"Cadet Zulka, understanding the tree, performing loose motions with hands in poket."*

One very fair cadet was playing the fool in the drill square, shamming, as they say in NDA. The NCO drill instructor walked up to him, held his belt and dressed him down. He then asked his name. The cadet would not give his name. After much bullshitting, he again asked his name. Fearing he would be awarded restrictions, the cadet gave his name with great reluctance.

"Cadet Hoffland, Saab."

The drill instructor's eyes lit up. *"Cadet Half land, accha isi lie naam batane mein sharama raha thaa."* (Cadet Half land, that's why you were shy of telling your name).

Come Again.......

- *Cadet Sergeant Major (CSM) -A Cadet appointment who runs the squadron. Normally a hard task master who is feared by the squadron.*

The cadets have a tremendous NDA spirit. Anything good that happens has to be ex NDA. One squadron had a brilliant, but a lost case surd CSM. One first termer received sweets sent to him by his mother. The first termer offered the sweets to the CSM. Quite surprised, the CSM asked him, "Who has sent these sweets?"

"My mother," replied the first termer.

"Good spirit. Is she also ex NDA?" asked the CSM seriously, then suddenly realized his 'faux pas' and went silent.

On another occasion, this CSM went to the Squady's office to give the daily report. The CSM read out the day's report. The Squady all this while was busy writing something, so probably did not register. He, then without looking up said, "Come again". The CSM saluted, turned about, smartly marched out of the office, turned around and asked, "May I come in sir?" and marched in again.

"What happened?" asked the Squady.

"Sir, you told me to come again," replied the CSM innocently.

The Squady smiled, and then noticing that the colour of CSM's turban was not the right shade, asked him, "I don't think you are wearing the right colour turban CSM. What colour is it?"

The CSM not quite sure, started craning his neck backwards, eyes rolled up to see his turban. Unable to see his own turban, he strained his neck upwards more and more till he staggered backwards.

The Squady smiled again, "There's a mirror in front of

you, use it before you fall and break your head."

It was quite like a dog going around in unending circles to catch his tail.

The Dirty Dozen

The cadets on arrival in the academy are issued a dozen handkerchiefs (hankies) for the entire period of three years. They are required to carry a clean hanky on person all the time. What is incredibly silly is that you are required to deposit them back before passing out. It is indeed ridiculous to expect the cadets to return all these hankies after three years. Many heads have rolled for want of a dozen hankies, including mine, for which I was de-striped as a fifth term sergeant.

What happens therefore is, as the end of term approaches, the passing out sixth termers put the sergeants on the job to collect hankies for them to return. There is a general drive by the sergeants to collect hankies from cadets for the sixth termers.

After the classes, the cadets are stopped and a hanky check parade is carried out. The cadets are well aware of this, so they take all precautions to keep their hankies clean to avoid losing them to sergeants. Some carry two, one for use and the other for inspection. Some keep them in a polythene cover to keep them clean. None of that, however, works with the sergeants who have their aim crystal clear.

Some lose their hankies because they are actually dirty. Others lose them because they are clean. Some lose theirs for carrying them in a polythene bag. They are told "What do you think the hankies are meant for?" Some are confiscated because they still have the trade mark label on it, and some for the absence of it. The hankies will go whatever you try. You of course cannot take the chance of not carrying one. If you did that, you could be reporting to the sergeant, with, may be a dozen hankies and lose them all. Losing one is better than the buggery and losing twelve. Those carrying two, land up in a decision dilemma of which one to lose, the new or the old. If they produce the new one, it just might get through if the sergeant has collected his quota for the day. The old one is gone for sure. The beauty is, every term all hankies are returned by the sixth termers and no one pays for any.

As the operation comes to an end, one can see the sense of satisfaction on the sergeant's face, quite like a fisherman with a good catch, a bulging satchel full of confiscated hankies. He will then give a small moral lecture, making others believe they were the guilty party and walk away as if nothing has happened.

Like the seasons which never fail to arrive, the ritual 'dirty dozen' comes every term without fail. And like matter which can neither be created nor destroyed, it only changes form; the hankies in NDA are never purchased by anyone, they only change hands, from juniors to seniors. As they say, every dog has his day.

One wonders what will happen to the last course, if ever the NDA was to close down. One thing I am certain, no cadet is going to pay for the hankies, they will manage, somehow.

The Breakfast Break

The most sought after meal in the academy is the breakfast. It is a sumptuous meal consisting of a lot of stuff. And precisely it is this meal that the sergeants ensure that you don't eat in peace, while they do. The sergeants are permitted to leave the classes a few minutes before the break to take position on the road to ensure strict cycle discipline. The cadets, on the other hand, try to beat the sergeants by all kinds of tricks. Some try to pill their way through. Pill is a NDA term to bluff your way through. Some successfully bluff the civilian instructors by saying that they have to go and get a tank issued for the military class. Others turn their corporal stripes upside down and say that they are special sergeants and have to take post on the road. Such tricks succeed only with the new civilian staff. When they do succeed to get out of the class early, they would take off on their bikes at break-neck speed to clear the area before the sergeants can take post.

Quite often they would be surprised by the sergeant leaping out from behind a tree at the nick of time. And then, it would be 'what a fall my countrymen', to see them all come crashing down on top of each other. Then would follow the usual drill of cycle riding the cadets for the entire breakfast break. And it would be a case of an egg and a toast, so near yet too far.

The first and second termers have a double outdoor almost every day. This means both PT and Drill periods. Sometimes all three periods are outdoors, which means they are a smelling, filthy lump of shit. Thus everyday they need to bathe. On the other hand the third and fifth terms called cool terms, have a lot of 'Flats'. Flats are the days when a course does not have PT or Drill classes but only academic classes. On these days the cadets wear normal working dress, as such do not need to bathe before breakfast and walk straight to the mess. The first and second termers thus are never the first ones in the mess.

The first termers those days were not permitted to use bicycles. They were required to run from one place to another. Even if they got to the mess in time at all, they would only be too lucky to get anything to eat. Some would end up doing 'Blow hot, blow cold', that is drinking hot coffee and ice cold water alternately. The more lucky ones would get an option to finish their porridge with a fork, after which they could eat the way they liked; only the porridge with a fork would never finish. If the first termer found a few seconds of solitude, he would first pack up a few toasts under his shirt and then proceed with eating. A cold toast in the shirt is better than ten in the plate with a second termer in front.

Porridge, eggs to order, ham and bacon, milk, tea and coffee and unlimited toasts. You name it and it was there. The NDA breakfast was a feast because it followed a couple of outdoor periods that could be PT, drill or swimming that leave you tired and hungry. A dozen palm sized toasts per cadet on outdoor days was the norm, four with the porridge (privilege allowed only fourth term onwards), four with eggs and the balance with a whiff of butter and jam, an art mastered over time.

Sadism thy name is 'NDA Sergeant'. It was the effort of an average sergeant to deny you a decent breakfast. The most miserable in a place of unequals, were the poor first termers.

The first termers wore civilian clothes for first few weeks till their uniforms were stitched and issued. Adding to the misery of the civil dress in the summers was the tie. A tie which spent more time rolled up in the first termer's mouth than hanging down his deflated chest.

First termer, if he did manage to dodge his way to the cadets' mess, was often a saga of 'so near- yet so far'. Six inches from ham, bacon and bread, he sits with a mouth full of rolled up tie. If it was not his tie, he would be gulping ice cold water and hot coffee alternately, glass after glass.

Bathing in breakfast break was compulsory. It was a big time waster. In the community bathing, when time was precious, first termer was the last to get a chance to get under the shower. They had to wait till the seniors were done. Some skipped bath to save precious time, simply wetting their hairs to get past suspicious seniors.

The cadets mastered the art of quick dressing. A course mate, Ajay Tanwar of Lima Squadron had a great experience. After a lightening bath, he quickly changed to working dress which was a well starched khaki shirt, a knee length K D short, stockings and garter flashes.

The KD shorts were well starched and knee length. It was not possible for a cadet to pull his shirt tight himself without spoiling the crease. Four civilian orderlies would sit on their haunches in a row in the central lobby to pull the shirts of the cadets. The cadets would rush down the three floors, stand facing the orderly who would shove his hands up the thighs, grab the shirt and pull it down. You would then turn around facing your backside towards his face. He would do likewise, shove his hands up your thighs and pull the shirt down, that removed the wrinkles and made you look smart and off you went to the mess as fast as you could.

Ajay Tanwar sprinted down the stairs and stood in front of an orderly named Banwari. In a single well rehearsed motion, mastered over the years, Banwari shoved his hands upwards, grabbed and pulled down hard.

Ajay Tanwar screamed and jumped three feet in the air as Banwari fell back on his bumps, scared to death. Banwari had grabbed his family jewels and pulled them down hard along with the shirt.

Ajay Tanwar, in a hurry for breakfast had forgotten to wear his underwear. He just about survived castration.

The many enemies of Breakfast break in the degree of nuisance value were:

Sergeants. Single minded devotion to deny others a decent

breakfast, while they will have theirs in peace. The fifth term had no double outdoors before the breakfast break and often not even one. They were thus mostly not required to bathe during the breakfast break. They were allowed to leave the class early and join late and had all the time in the world.

Academy Instructors. Thought the best time to take extra few minutes was the breakfast break. The cadets were hardly ever interested, let alone during the breakfast break.

Fourth Termers. Were the divisional seniors of the first termers. Insisted on conducting a bathing inspection daily during the breakfast break after he was ready, leaving you no time. No cycle allowed from squadron to the Mess. One had to walk up and down and manage the sergeants on post en route.

No Satchels Allowed in the Mess. The satchel had to be left in the squadron. This meant that after breakfast you had to come back to the squadron, run up one or two floors to collect the satchel from your cabin, run down back to battalion quadrangle where the cycle stands were, and finally clear the pre-determined point fifteen minutes before the start of classes. All this gave you very little time indeed. In the best of times, it was not enough. With all the obstacles thrown in and a long uphill cycling, it was almost impossible, but the cadets managed somehow by their innovations.

Sarad Chandra was a good friend. A sincere and a hard working fellow from Kerala. However hard I tried, he always beat me to the breakfast table, duly bathed and powdered.

One day I asked him, "How do you manage to come to the mess so fast every time, what is your secret mate?"

"Do you sing a song while bathing and dressing up?" he asked.

"No," I said.

"Do that and you'll start doing everything faster," he replied. Good idea, I thought. I tried that, yet I could not catch up with him. I again told him it doesn't work man.

"Which song do you sing?" he asked.

"A ghazal of Talat Mahmood," I replied.

"How can that work, ghazals are too slow. Rather than help, it will slow you down. You know which song I sing?" he asked. "No, which one," I asked. He then sang a few lines of a Malayalam song. It was a peppy sixty rpm song, being sung at ninety rpm. It was a real fast number on which only a belly dancer could dance.

He said, "I sing this song and bathe and change at this pace. Find yourself a hot and fast number and see the difference."

After much research, I found one on which the famous film cabaret dancer Helen had danced in a film.

Believe me, it worked. I bathed and changed at the pace of the song and soon caught up with Sarad. My breakfast consumption shot up by fifty percent. I've ever been grateful to this friend in need.

Lesson learnt: *'Complicated problems have simple solutions'*.

The Academy Auditorium

One of the most interesting places in the academy, which no cadet ever forgets is the academy auditorium. All addresses, movies and cultural programmes are run there. The hall is an impressive three floor building. The first floor balcony is meant for the officers and families, the ground floor and the second floor balcony for the cadets. The second floor i.e. the top floor balcony is dedicated to the first termers. I suppose, far from gaze of everyone, first termers either courageous or foolish enough to go for the film are entirely at the disposal of the sergeants. The first termers do everything except watch a film.

It is a common sight to watch the first termers rolling around in different directions when the film is on, or watching the film from either hands down or murga (Rooster) position, literally getting the proverbial bird's eye view of the film. Nonetheless, the first termers watch the film, no matter what position they are in. Those who are lucky enough to be on their seats, watch the film sitting in attention, arms stretched.

Actual drama starts after the film is over. The Director of the show is the Academy Cadet Adjutant (ACA). ACA is a prestigious appointment. He literally runs the show and ensures academy discipline. Somehow the ACA is always a poker faced bloke who has vowed never to smile and walks with the arrogance of a sultan. The post film drama would start with a moral lecture from the ACA, telling the academy how useless they all were and how unacceptable and deplorable was the discipline. While he did this, the sergeants would take ambush positions and seal all the escape routes. Then followed the long trek to the cadets' mess either rolling down, or hunching, or, if you are lucky just sprinting down to the mess as if your tail was on fire. En route, you would come across sergeants emerging from the shadows to grab you if you are not sprinting faster than Husain Bolt. You were in for a tough time if you happened to have a fourth termer in your squad. The fourth termers traditionally do not get along with the sergeants. They think they are too senior to obey the sergeants without resentment and the sergeants would specifically target the fourth termer. As the junior, you were the one at the head of the squad. You would invariably get caught in the cross fire between the two, the sergeant yelling to step out and the fourth termer asking you to slow down. One way or the other, it had the potential to ruin the weekend.

Nonetheless, it is interesting that despite the itinerary of the auditorium, it was crowded every film day. Cadets would not mind watching the film hands down for three hours,or facing backwards in murga position, watching the film upside down.

Cadets were on the look out to take off and disappear in the darkness to avoid the painful journey to the mess. When they did, the sergeant would be in a dilemma to chase or not to chase. If he did, the others would disappear too and there was no guarantee he would catch the defaulter, so a well calculated risk. As for me, I still carry a scar from that journey.

Well, some gutsy guy took that chance with the ACA. The ACA was known more for his brains than brawns. The cadet had calculated well and appreciated that even if the ACA decided to give a chase, he would out run him and disappear in the darkness. The ACA stopped his squad of six cadets and got them rolling. The moment ACA looked the other way, he took off. Surprised, the ACA gave chase, and seeing their chance, the rest also bolted away. The ACA decided to continue the chase, knowing he could do little about the other five. He was determined to catch this 'son of a b***h' who dared challenge the ACA. Then started the cat and mouse chase. Running all over the place with the ACA in pursuit, the cadet entered the battalion quadrangle of the second battalion, hoping to shake him off. By then the ACA was beginning to tire and the distance between the two began to increase. The chase continued around the cycle stands as the ACA threatened him with dire consequences if he did not stop. He had taken too many chances and would be a fool to stop now. Both of them went around in circles a number of times. Suddenly, the escapee was stunned by a dhap on his face. Taken aback, he fell to the ground and the ACA wrestled to pin him down.

What had happened was, after a few rounds, the ACA was half dead. He accepted the fate and gave up the chase. Holding his aching stomach, his heart threatening to explode, he leaned against the cycle rack to catch his breath. Then he heard the footsteps approaching, getting louder and louder. When his moment came, he sprang the ambush. The escapee had not realized that the ACA had given up the chase and had halted. Unaware, he kept running around in an auto mode passing the point where the ACA was recovering his breath and by default had laid the ambush. What happened then is history.

NDA Auditorium

NDA Punishments

The NDA, over a period of time has evolved a variety of punishments, which have one single aim, that is to make life as difficult as possible for the juniors. Most of these are applied on the first termers though. Majority of these punishments are physical in nature, and build up the cadets physically and mentally. Well the jury will always be divided on the utility of these punishments and ragging. The fact remains that the cadets, in the earlier days, owed their physical endurance to the seniors who put them through a tough physical regime through the so called ragging. The cadets grew up in an environment where cribbing was considered un-officer like. The NDA cadets thus found it easy to cope with the tough military life with that wonderful attitude. In fact, all through the six terms, the cadets had only one quotation written on the desk "Cribbing is un-officer like" and they followed it in letter and spirit all through their career.

Ragging has been banned in NDA since long. NDA punishments in vogue those days are described below. You will understand their impact better if you try doing some of them, for fun sake.

Puttee parade: It consisted of making the cadets change various dresses at a rather rapid pace one after the other. The senior would spell out the time and dress e.g. riding rig, one minute. The worst part of this punishment, was, that it screwed up your whole cabin and it was a bigger punishment to reset the whole thing in place again. Apart from the conventional dresses, some interesting ones that were part of the puttee parade are described ahead. There are always two sides to a coin. The puttee parades trained you to dress up fast and save time. That was a big asset in the breakfast break, for in the NDA, time is bread and butter.

Birthday suit : As the name suggests it consisted of no dress at all. The cadets were required to appear all nude. It was a funny sight to see tens of crew cut cadets of all sizes and hues running around in birthday suit.

Phantom rig : A cadet would wear an overall with a swimming costume on top, covering face with a woollen socks.

Mummy rig : A cadet was required to cover himself with riding puttees all over from toe to head, leaving just enough slit to see through. It needed the help of others to dress up in this rig. It became difficult to walk as you would be all stiff.

Multi pack bajri order: A cadet would wear specified number of packs filled with sand and stones. Sometimes you were told to report with twenty packs. Obviously, a cadet could not carry all these on his back, so he would be having three on his back and dragging the rest behind him. Like a queen, there would be a procession of cadets behind him helping with the other packs. A cadet has only one pack issued to him. He would therefore borrow the rest from course mates and spend long time filling them with sand. Imagine filling twenty packs with bajri.

Seventh heaven: The NDA cabins have a wire mesh on the front side from top of the door till the roof. So, really speaking, any one could peep inside the cabin. It had, if I remember right, about fifteen horizontal wires. This punishment required the cadet to jump and hang on to the seventh wire till you were told to come down. The mesh wires were rather thin and hanging on for anything longer than a few seconds was real torture. Clearly it was far from a feeling of being in the seventh heaven; it was much closer to hell.

Rotation : A cadet was asked to bend down from the waist, touch the floor with his index finger and rotate around it like a top. It sent one into dizzy in no time. The cadets tried to beat the punishment by rotating in opposite direction alternately, when the senior was looking away, which was not often.

Rolling : The most popular of all the punishments and generally speaking, the easier one to handle. Simple, just roll like a ball. Things got bad when someone decided to exercise the variants. These could be rolling down the staircase, rolling under corridor carpet, and side rolling etc. Rolling bareback under a corridor mat was no fun either.

Heat treatment : By far the worst and most dreaded of all the punishments. It had many variants from standing bare feet on the sizzling tarmac in the peak of summers to rolling bare back or crawling. Quite painful and big blisters on your hands or feet were invariably the result.

Bicycles: Bicycles often rode the cadets rather than the other way round. Most Saturday afternoons were dedicated to academy "Screw parade". On Saturday afternoons, the sergeants would leave the classes earlier than usual to close all the escape routes. The road from Ashoka Pillar to Pashaan would be full of cadets with the cycles over their heads. A great sight indeed, over a thousand cadets with bikes on their heads. Some would be walking on their knees with the bikes on their shoulders. Often the cadets preferred not to take their bikes on Saturdays and run down to the classes instead. Chances were that you could get stuck in a worse predicament than with the bikes.

The exercise was normally undertaken when the ACA felt that the discipline in the academy was slackening. It was also the NDA way of starting your weekend. It all depended on the mood of the ACA and that was rarely good, if ever, and it was worst on Saturdays.

Mug parade: Was one of the most dreaded ragging punishments. The first termers were lined up with their steel enamelled mugs in the hand. On instruction from the circus master, invariably the sadist fourth term corporal, the cadet at the end of the line would bang the head of the one standing in front of him, and so it would continue down the line. When the whole line was through, they would turn around and repeat performance would follow. If you showed sympathy and hit softly, you had it. The senior would call you and give you a real hard one himself with the mug.

At the end of the exercise, everyone landed up with big bumps on either side of their heads resembling large headphones. To add insult to injury, this manoeuvre was invariably executed during the tea time, when others enjoyed tea. The first termer never hoped to get tea in the first term.

Stool parade: A bed side stool/peg table was inverted with legs facing upwards. A cadet was asked to remove his shoes. He was then told to mount up on the legs of the stool and assume the traditional "murga" (Rooster) position. In this position you are required to get on your haunches and hold your ears from under your thighs. A tough position as the stool legs painfully dig into the sole of the feet. Balancing on the stool legs itself is no mean task.

Rolling under shower: Involved rolling under shower in the bathroom, with alternate cold and hot water shower running.

Blow Hot Blow Cold: Privilege exercised by a senior in the cadets' mess. A cadet would be asked to drain a glass of hot coffee and ice cold water alternately.

Stroll under the Dining table: Reserved for first termers. He would be asked to get under the dining table and remove the shoe lace of somebody at the other end of the long table. Rarely did anyone, if ever, succeed. The cadet under the table became a football and got booted from all directions.

Academy General Knowledge

The first termers were required to learn the 'Academy General Knowledge' as given below :-

Sudan Block

- Formation Stone - 6th Oct 1949
- Shape - Field Gun
- Cost of Building - 2.84 Crores
- Number of pillars in Sudan Block - 44
- Height of Sudan Block - 132 ft

- Steps in front of Sudan Block - 26

Science Block

- Shape - Machine Gun
- Diameter of the dome - 9 ft
- Height of Block - 110 ft

Drill Square- Khetrapal Parade Ground

(Named after Second Lieutenant Arun Khetrapal, Param Vir Chakra {Posthumous}, 38th Course 'F' Squadron, winner of the highest gallantry award)

- Dimension - 450 x 150 m
- Mast in Drill Square belongs to - INS Delhi
- Height of Mast at Drill Square - 99 ¾ ft
- Perimeter of Drill Square - 1.2 km

Hut of Remembrance

- Time period of making - Completed in May 1957
- Constructed by cadets of 10th to 17th Course
- Unveiled by - Gen KS Thimayya, Chief of the Army Staff, on 1st June 1957
- Area of hut - 2.73 Acres

Squadron Data Sheet

- Number of tiles in cabin - 144
- Number of cadets in a squadron - 124
- Number of bathrooms in a squadron - 6
- Length of corridors - 52 ½ front rolls
- Length of flanks - 23 ½ front rolls
- Seventh Heaven - Seventh strand of wire mesh in

		the cadet's cabin. Used to hang the first termers

Cadets' Mess

• Number of seats in Cadets' Mess	-	2111
• Number of bulbs in Cadets' Mess	-	252
• Steps in front of Mess	-	10
• Glass used in Cadets' Mess	-	Originates from Belgium
• Wood used in Cadets' Mess	-	Originates from Burma

Miscellaneous

• Fastest bike in Academy	-	F-16 (Foxtrot Squadron, Bike No 16)
• Highest bike in Academy	-	K-2
• Heaviest bike	-	K-100
• Who are Pongos	-	Army cadets
• Academy mystery	-	Where are 2nd termers?
• Academy disgust	-	Cabin cupboard
• Academy wonder	-	Finding an empty pen refill
• Name of Rape hill	-	Table Top
• Academy D**k	-	Ask an ex NDA
• Commandant's b**ls	-	Big stone balls in front of Quarter Master (QM) fort
• Academy question	-	Which is the next Lecture Hall?
• Academy answer	-	Do not know
• Shape of library	-	Anchor of a ship

- Academy garland - Bike
- Academy colour - Maroon
- Academy joke - NDA makes boys into men
- Academy 16th Squadron - Military Hospital
- Academy Jurassic Park - Number 4 Battalion
- NDA Day - 16th Jan
- Academy Waterloo - 10 m Jump
- Helen's **** (Ask an ex NDA) - ETT Gate (Equitation Lines)
- Academy condom - Wind sock of Air Force Training Team (AFTT)
- Academy kela - Cadet coming 16th in X- Country (Only first fifteen get a medal)
- Academy full stop - Last in X-country
- Academy vibrator - Peacock bay
- Academy prostitute - 4th termers
- Academy pillar - Safety pin
- Academy pondy - "The Guide" (English text book for 1st termers by RK NARAYAN)
- Height of Ashoka Pillar - 29 ft
- Drawing of Habibullah-hall - Lila Books' 36 Paintings

INDIAN MILITARY ACADEMY

CODE OF THE WARRIOR

"I Am A Warrior
Defending My Nation Is My Dharma."

I Will

* Train My Mind, Body and Spirit to Fight.
* Excel in all Devices and Weapons of War - Present and Future.
* Always Protect the Weak.
* Be Truthful and Forthright.
* Be Humane, Cultured and Compassionate.
* Fight and Embrace the Consequences Willingly.

"God Give me Strength that
I ask Nothing of You"

GLORIOUS CHETWODE & DRILL SQUARE

THE MAKING OF A WARRIOR

LEST WE FORGET

FREEDOM DOES NOT COME CHEAP :
BLOOD GUTS AND GLORY

PART III

INDIAN MILITARY ACADEMY

Chetwode Motto

*"The Safety, Honour and Welfare of your Country
come first always and every time.
The Honour, Welfare and Comfort of the Men
you command come next.
Your own Ease, Comfort and Safety
come last always and every time."*

Sir Philip Chetwode

The Genesis

The **Indian Military Academy (IMA), located at Dehradun** is the premier officer training institution of the Indian Army. It trains officers for the army. There are similar academies for the other services.

In 1932, the Indian Military College was established outside of Dehradun by the Prince of Wales as a feeder school for Indians headed to the Royal Military Academy at Sandhurst, England.

The Montague-Chelmsford Reforms enabled ten Indians to be sent to Sandhurst for training. Later the Round Table Conference in London in 1930 recommended the establishment of an Indian version of Sandhurst. In July 1931 a committee under the chairmanship of Field Marshal Sir Philip Chetwode, then Commander-in-Chief, India, recommended setting up an academy for training forty entrants for a term of six months.

The Academy became functional from 1 October 1932 with a course strength of 40 Gentlemen Cadets. Brigadier L P Collins, DSO, OBE was the first Commandant. The first course had on its roll Sam Manekshaw, Smith Dun and Musa Khan. All of them later became the chiefs of the armies of their respective countries, namely India, Burma and Pakistan. The course was christened 'PIONEERS'. The government acquired the estate of the erstwhile Railway College at Dehradun, which had the appropriate buildings and a fairly extensive campus to meet the requirements of the Academy at its birth.

Formal inauguration of the Academy was done at the end of the first term on 10 December 1932. Field Marshal Sir Philip Chetwode, Baronet GCB, OM, GCSI, KCMG, DSO, then Commander-in-Chief of India, after whom the main building and its central hall are named, inaugurated the Academy.

The highlight was the inaugural address by Sir Philip Chetwode, which was delivered in the hall now named after him. A passage from his address has been adopted as the Credo of the Academy. This is known as the 'Chetwode Motto' and is the motto of the officers of the Indian Army.

After independence Brigadier Thakur Mahadeo Singh, DSO was the first Indian Commandant. At the time of independence the movable property of the academy were divided between India and Pakistan. The Gentlemen Cadets, who opted to go to Pakistan, left the academy on the night of 14 October 1947. The first two generations of the officers of the Pakistan Army were products of the Indian Military Academy.

During the Second World War, the inter-dependence between the three services had been realized. The Government of India, therefore, approved the creation of an Inter-Services Wing for training cadets for the Army, the Navy and the Air Force together. Thus, in January 1949, the Academy was redesignated the Armed Forces Academy, with the military wing remaining on the present campus at Prem Nagar and the Inter-Services Wing established at Clement Town.

In consonance with national aspirations, the Indian Military Academy was redesignated the National Defence Academy in

January 1950. The Inter-Services Wing became Joint Services Wing (JSW). The first JSW Course passed out in December 1950.

In December 1954, the Joint Services Wing moved to an entirely new campus at Khadakwasla, and along with it went its name, formation sign and the Commandant. The Indian Military Academy (then renamed the Military College) recovered its original identity and role.

In 1960, the Military College was renamed the Indian Military Academy.

IMA is located in the serene surroundings of Dehradun on the bank of river Tons. It has produced over 70,000 well trained and motivated officers who have been the sentinels of the country. A look at the memorial in IMA with its unending list of martyrs speaks of the sacrifice by the officers who have passed out of this wonderful academy. Unlike National Defence Academy (NDA), Khadakwasla, IMA is not on a dedicated piece of land. On either sides of IMA lie the shanty shops along the Chakrata road that runs right through the centre of the academy, and worse, right in front of the Chetwode Parade Ground, the most sacred place in IMA. For every passing out parade, the road has to be closed. On routine days, the civil vehicular traffic has to be regulated to allow the cadets to cross the road causing inconvenience to all. It is time the issue is resolved either by construction of tunnel or a fly over or whichever means. The premier academy must have a personality that should leave an indelible positive impression, the way NDA does. At present it does not and a lot needs to be done. It should be a case of love at first sight.

Unlike NDA, barring the main blocks, named after the first two Commandants, Brigadiers Collins and Kingsley, the living barracks looked closer to stables than cadets' living during the period we did our training in 1978-79. For the ex NDAs, IMA was a big come down in every way. The days at IMA were not so long. One did not count DLTGH (Days left to go home) simply because life was far more comfortable mentally and physically.

And before you realized, the year was gone and it was time to walk over *"The Last Step"*. Nevertheless one gets used to the ways of IMA soon. A lot has happened in IMA since then and infrastructure has improved immensely. The environment remains more or less the same, as it is bound to.

The cadets in IMA are called Gentlemen Cadets, GC for short. For from here, they become officers. Gentlemanliness is intrinsic to an army officer and a good military leader. The academy has a peculiar military environment. The uniform is 'Olive Greens' and the cadets' living accommodation was typically old military type barracks. The cadets would be running around with rifles doing weapon training, bayonet fighting , rifle drill or battle obstacle course. Unlike the NDA, IMA has cadets coming from the NDA, the Army Cadet College (ACC), the Direct Entries (DE) and of course the Technical Entries, popularly known as the TD or the Techy Dopes. A word on each of these will help to understand the IMA environment better.

The Technical Entry called Techy Dopes (TD for short): The technical entry are the cadets who join the academy after doing their technical graduation and then go on to join the technical services in the army. The techies, as they are called, are of different 'models' based on their mode of entry. The traditional variety join the IMA after Engineering graduation. The next variety are those selected halfway from the engineering colleges called the Unit Entry Scheme (UES). On completion of technical graduation they join the IMA. The last variety is picked up after school. They are trained at IMA for one year, put through technical graduation at different academies and pass out from the IMA. They are called the Technical Entry Scheme (TES). So a real cosmopolitan collection as you can see.

The candidates for the army are hard to come by, corporate world drawing the youth. The techies seem to know the open secret that the IMA will not kick them out except on moral and integrity grounds. So why bother too much. As long as they can survive twelve months, the President will be pleased to grant them commission.

The variety of entries make the IMA a cosmopolitan but an immensely interesting place. Of all the entries, the techy dopes carry the maximum potential to provide entertainment to the environment. Since they join the IMA after their post graduation, they are normally in their mid/late twenties and quite a few married with children. Sometime they misinterpret the terms and conditions and think they already have become commissioned officers when they join the IMA. Some will arrive in the academy looking for their plush offices. The reality dawns when the first senior whom they encounter tells them that their office is the mother earth and their job, front rolling for the next six months. He then goes rolling around to find his elusive office.

The Direct Entry (DE for short): The DEs join the academy after graduation and do year and six months training period. Coming from the college, their ideas are a bit on the wrong side having been loose cannons in the college with little or no discipline. Physically they are mostly under par. They derive their name from the belief in the academy that many of them would never get through the training if left to themselves. They have to be pushed through, hence the sobriquet 'Dhakka Entry' or 'Pushed Entry', if loosely translated in English. Generally of the 'live and let live' category, they are a harmless variety and are rarely seen as appointments beyond a rotational sergeant, quite contented to sail through without creating ripples. The DE is generally considered a lost variety that lives by the day and seems to realize he is in the wrong place.

The Army Cadet College (ACC) Entry: Is one where the deserving serving soldiers after going through the selection process join the ACC, Dehradun. They carry out initial three years training at the ACC located adjacent to the IMA, where they obtain a graduate degree and do the basic military training. Then they join the IMA for one year specialist military training and are commissioned into all arms and services. Many officers from this category have excelled and done the army proud.

Well, this category of cadets coming from the ranks do not have the advantage of ex NDAs where you go through a top class grooming. Being in service they were the only category that got their pay, and compared to the others who only got a paltry pocket money, had the money to fool around. English was not their forte and you could make out an ACC cadet from his accent and comparatively grown up and rugged looks. Generally quite contented with what they had achieved and being older, they looked forward to getting commissioned into the services, the softer option.

The Ex NDA: Cadets carry a reputation of being crooks of the highest order, and for giving a convincing excuse even for a broad daylight murder. Can't blame them, it's all about survival in that hostile place called NDA. Youngest of all the variety, they do well and hold all important appointments and call the shots in the academy. The ex NDA also undergo training for one year at IMA after three years at NDA where they do their graduation.

So, in the second term all the three categories of entries described above come together to make that heady cocktail.

IMA as it was

When I joined the IMA after NDA in June 78, the famous Major John Taylor from Gurkhas was the academy Adjutant. He had a particular liking for the NDA cadets, a kind of love and hate relationship. We had heard a lot about him and were looking forward to seeing him. He was known to have a tremendous sense of humour. The first itinerary was the Adjutant's address to the mixed bag course. We sat impatiently waiting for the man about whom we had heard so much.

In strode a Major, slightly on the heavier side with a bit of a paunch, sporting a real crew cut, typical of Gurkha officers. Surely he couldn't be the Adjutant of IMA, I thought. We were used to the 'tall-dark-handsome' variety of Adjutants in the NDA from Brigade of The Guards. That was a great disappointment I thought. He stood in the centre of the stage looking at the cadets, saying nothing. As the hall slowly fell silent, he spoke. Well, it has been a long time but I can remember some of what he said to us on that first working day at IMA.

"If you have finished, may I have your permission to begin gentlemen." The hall fell pin drop silent. Major John Taylor had a squeaky voice which did not go well either with his appointment or with his personality. "I am Taylor, they call me John Taylor. I am the Adjutant IMA. I have little doubts that you have heard of me already, and now you have the privilege of seeing me. I am not a tailor by profession but you will soon realize how important are both at the IMA, the tailor with an 'I' and a Taylor with a 'Y'. We shall hereafter have a long association of one year, which I am delighted to say, is not going to be a very pleasant one. After one year some will have pleasant memories and some not. Going by my looks and personality, some of you from the NDA would have hoped to see a more handsome man. I am sorry to disappoint you. To clarify some doubts in your minds, let me tell you, I am no relation of Elizabeth Taylor. I don't take pride in the similarity of our names, I am told she does."

"Remember boys, you all have now graduated from Cadets to Gentleman Cadets. We expect you to behave like one. My friends from the ACC, grow up from where you have come and stop bloody singing *'mein toh sala sahib ban gaya'*, (Wow, I have become a big man, a song from Indian film) and get down to business. My friends the DE, my entry, you have been here for six months, it is time to get going on your own, you don't need dhakka (push) to pass out, do you?"

"And to the boys from the NDA, my favourite entry. Grow up, remember, now you have graduated from shorts to pants." And then raising his voice "Don't try your monkey tricks with me, for these are the hands that have rocked the cradle." (NDA is known as 'The Cradle of Leadership').

John T, as he was popularly known, personified IMA to the cadets. Their world revolved around him, mostly. He was always there in the parade ground, sometimes on his own feet and sometimes on the horse back. When he was there, the cadets were rarely on their feet, they were mostly on the move at one hundred and eighty steps per minute.

The Adjutant is the cog around which the wheel of the academy runs, yet his most prominent and visible commitment is on the parade ground.

It would be in fitness of things to begin with some wonderful incidents of the IMA drill square.

Before I do that, a mention of my first close encounter with John T would be in order. As the Academy Under Officer (AUO) in IMA, I would report to him for orders once in a while after the classes. One day he looked at me, "Sengar, I can see that you are sporting some hair above your upper lip, underlining your nose, which you may be calling a moustache. But do you know what it looks like? It looks like a bloody rat's tail." He was right, there were a few strands then. I have done much better since. And then he says, "What arm have you opted for?" "Infantry, Brigade of the GUARDS," I replied. "Oh, you will be a Guardsman. So you will have a missile here, here and here (pointing at the beret, right breast pocket and then towards the backside)." We became good friends after that. Unfortunately never had the privilege of meeting him after commissioning.

How Tall Are You

IMA prides itself with one of the finest standards of drill anywhere in the world. The standards are show cased in its full glory during the passing out parade.

The parade is watched by Generals, serving and retired- Indian and foreign, parents of passing out cadets and diplomats etc. That is the most glamorous part of IMA passing out ceremony. A lot of hard work goes into rehearsing and preparing for the passing out parade. The rehearsals start almost four weeks in advance and are supervised personally by the adjutant. It is his show.

One day the academy was rehearsing the passing out parade. Cadet Ahluwaliah, a very tall lanky surd, generally uncoordinated in movement and gait was spoiling the drill of the whole company. John T who had been quietly watching Ahluwaliah for quite some time cantered up to him on his horse. He brought his horse within inches of Ahluwalia's face. The horse's frothy mouth and strong breathing sprayed the cadet's face as he fought his reaction to back off, worse, the horse might just bite him, he feared. A cadet in IMA has little choice but to stand rock still, no matter the consequences. John T roared in his squeaky voice,

"How is the weather up there surdy boy? Do you know you are the most cock-eyed disjointed son of a gun that ever set foot in IMA? My dear boy do you realize that you are screwing up the drill of the whole academy. How god damn tall are you?"

"Six feet four inches sir," yelled Ahluwaliah.

"Six feet four inches! Never seen shit piled that high in my life," yelled back John T and quietly trotted away.

The academy burst out laughing. Ahluwaliah was relieved to get the horse off his face, the rest did not matter.

The Horse Sense

Gentlemen Cadets take immense pride in putting up a good show on passing out parade and practice hard for it. There are days when they just want a break. Like any other thing, there are days when nothing goes right however hard you might try. During one such day when John T was present for the rehearsal, nothing seemed to be going right. The turns were atrocious, swinging and digging terrible and rifle drill disappointing. In spite of his best efforts to egg the cadets on, at which he was good, nothing improved.

Thinking probably a change in word of command will make the difference, he told the drill Subedar Major to give the word of command for *'Salami Shastra'* (salute with arms). Typical of the drill Subedar Major, he egged on the cadets and after a brief pep talk gave a full throated word of command *'Salami Shastra'*. The cadets went through the drill motion. Just as luck would have it, it was worse than before. There was a lousy rumbling sound of thumping of rifle butts and stamping of feet as opposed to crisp crack of front guard and slapping of the rifle butts. There was pin-drop silence as cadets waited for John T's reaction. As moments passed, the Adjutant sat quietly on his horse, head bowed down in disappointment. Just as the drill movement finished, Adjutant's horse shat a wholesome, accompanied by a mighty fart that was heard way off. John T was not the one to miss such an opportunity. He rose and stood tall on the stirrups, turned fully backwards from his waist, looked at the rear of the horse, turned around, surveyed the parade from left to right and yelled,

"Look-look, like me, that is what even my horse thinks of your drill. It is all bloody shit."

John T's Drill Test

Like the NDA, the drill test again comes to haunt the cadets at IMA. Here too passing the drill test gave the cadets the privilege to go to the town on weekends. Considered a great privilege, the GCs tried hard to pass the test as early as they possibly could.

John T conducted the drill test himself. But, his approach was, as a matter of fact, quite casual. John T enjoyed dealing with cadets. When he wanted to be entertained he would walk up to a group of techies or ex ACC cadets. When he was in real need of entertainment, he chose the ex NDAs.

John T while conducting the drill test almost never looked at the GC appearing in the test. He would stand with his back towards that cadet and keep talking to the others. It was disheartening to those who were good at drill and a boon for those who were not. All the while he would be on the lookout for the amount of sound produced by the thumping of the rifle butts. If he was satisfied with the sound, the cadet would invariably pass irrespective of the standard of his drill, if not, you were back to square one. To produce a big bang the cadets would loosen the nuts and bolts of the rifle. One Dhakka cadet, GC Cherian who was rather atrocious at drill thought it was the best bet to pass and lined up for the test with all his nuts and bolts loose.

The Drill NCO shouted

"Drill test shuru karne ki anumati chahta hoon sriman."

(Request permission to commence the drill test sir).

"Carry on," replied the Adjutant as he continued to bullshit the cadets.

"GC Salami Shastra," (Present arms) commanded the Drill NCO.

GC Cherian in full josh tossed the rifle to his side and banged his palm on the butt. As he slammed the butt, the metal butt plate went flying rattling on the tarmac.

John T turned slightly and glanced from the corner of his eyes and continued his chat with the waiting drill squad.

Unperturbed, Cherian continued with the drill. As he brought the rifle forward and slammed the front hand guard, the screws came off and the hand guard flew apart landing noisily on the floor.

John T raised his eye brows with a tilt of the head and continued his fun with the drill squad.

On the third motion, as he slammed his right hand on the rifle and stamped his boot, the magazine plate came off releasing the magazine spring like a rocket. It sailed over John T's shoulder and landed in front of him.

John T briefly surveyed the rifle parts lying all over without saying a word. He gave an expression of amusement and carried on as if nothing had happened.

"Baju Shastra," (Side arms) commanded the Drill instructor.

GC Cherian, by now sweating like a pig and breathing like a horse, slammed his right hand on what was left of the hand guard and stamped his foot.

As the luck would have it, this time the bayonet flew off. By now the rifle parts were scattered all around John T. GC Cherian stood there with a rifle that looked like a dressed chicken. Unperturbed he carried on as if nothing had happened.

As silence prevailed, the squad waited for something to happen. John T, who all this while had watched patiently, turned around and ambled close to Cherian. He stood facing him, surveyed the litter around and yelled.

"First it was the hand guard, then it was the butt plate, then it was the magazine and then it was the bayonet. Thank God there are no more commands or you would have exploded next. For your lousy drill you would never pass the test. You have done what has never been done in IMA before. You have shattered the myth of DRDO in many parts without firing a shot.

You have put DRDO to shame. And for your josh my boy,you have passed."

Then turning to the squad, "Will you please help him collect the remains of his rifle and bury them somewhere." And walked away.

John T was in a happy mood that day. GC Cherian was not the only one to benefit from his largesse.

There was a techy dope - GC Goswami, whose drill like many other things, was rather atrocious and had no hope in hell of passing the test till 'supper night carry on' came. It was his turn to appear for the test and he marched in, in his funny gait. The scenario was no different, John T bullshitting and talking to the drill squad facing the other way. A Drill NCO yelled the word of command *"Salami Shastra".* Goswami in his awkward style tossed up the rifle. The toss was so laborious and awful that the rifle almost fell from his hands. Goswami lunged at the rifle and just about managed to grab it. In the process his hand hit the bayonet and started bleeding. He grabbed his hand, forgot about the drill test and stopped.

John T hearing no further sound turned around and seeing his bleeding hand yelled, *"Good Lord Jesus Christ. Hum toh sirf pasina mangta hai, aur tum khoon bahata hai, pass, line tor."* (Good God, I only demand sweat and you give me blood, pass, break off).

Goswami could not believe his ears. He stood his ground waiting for he did not know what. John T repeated himself *"Engineer saab, pass. Ab jaiey".* (Mr Engineer, pass, now please move along). It was a day of lottery; the blood was a blessing in disguise. I guess a bit of blood can save a lot of sweat in peace.

Bee on the Arse

Mounted on his white charger like a war lord, John T was supervising his sultanate from a vantage point that gave a good view of the drill square. He suddenly kicked his horse and cantered away towards a squad in one end of the drill square.

He had seen GC KT Pundhir march with a funny swagger, his oversized hips shaking left to right. His right hand with half open palm was swinging around his hips in a poetic motion. John T couldn't digest a cat-walk in the drill square. He prodded his horse and stood right in front of Pundhir and said,

"KT, do you know what your drill is like. You have got the funniest bloody drill I have ever seen. When you walk swinging your sexy backside from side to side, and swing your arms around your arse with a cocked wrist, it looks as if a bee is sitting on your arse and (then swinging his right hand behind his back a couple of times in quick motion) you are telling it- get off my arse..., get off my arse.., get off my arse. I want no comedians here, get it. Now get hold of your backside and march like a Gentlemen Cadet," and trotted away bouncing on his white horse in search of another prey.

Havaldar Swamy; an Instructor Par Excellence

Well, while John T personified the IMA drill square when we were there, there is another character who needs a mention, a lesser mortal. He was Havaldar Swamy from the Engineers. A dark and tall, lean, mean sprightly drill instructor. A character straight from the US Marine Corps film. Dedication personified, he worked harder than the cadets on the drill square and sweated more. He changed his uniform for every period of drill in a day. He had had three tenures as drill instructor in IMA and constant drilling had turned his drill to almost robotic perfection. When he turned or marked time, his knees would touch his chest. When he swung, his arms would swing above shoulder levels both front and rear. When he marched, he would dig from above belt height denting the tarmac.

He was a monster on parade, strict on discipline and would not spare anyone. A drill period with him left many with swore feet and stress fractures. He would then be seen taking post to ensure that the roadside discipline is followed. He believed he owned the IMA and that was his commitment to the academy.

The training over and once you were commissioned, he would greet, wish and salute every commissioned cadet. A perfect example of a drill instructor.

DRILL INSTRUCTOR TRAINING

It's time to step out of the parade ground and see what happens elsewhere in the wonderful world of IMA.

A Techy at his Innovative Best

- *Directing Staff (DS) - Officer Instructor, usually a hard task master.*

The techy dopes often surprised you with their innovations. They were rather too technically qualified for a place like IMA, where you were drilled in the basics of soldiering and leadership. In such a place, an engineering degree was of little consequence and they were mostly a square peg in a round hole.

During a second term camp at the IMA, we had taken up platoon defence on the foothills of Mussoorie. It was the month of November and poured like hell the whole night. All our trenches were overflowing and communication trenches became small canals. The area looked like an unending lake with knee deep water everywhere. Everything was simply in utter chaos.

Totally drenched and chilled to the bone, our teeth rattled uncontrollably like a rattle snake. We tried to last as long as we could. Finally, not able to last any longer we decided to take shelter in a cowshed a few hundred yards away.

We could not have left our rifles behind. But our rifles were nowhere to be found. Where the hell have the rifles disappeared, we wondered, growing worried with every passing moment.

In the knee deep waters in a pitch dark night, we splashed around for our rifles without success. Tried deep sea diving in the hope of fishing them out from the fire trenches, to no avail. Hours later, we gave up, resolved to look for them the next morning. We wondered our fate if we failed to find them before the D S returned. Losing a weapon was a serious offence and meant certain relegation, if not a possible withdrawal.

When the day broke, the rain had eased a bit, but it was still drizzling pretty heavily. We returned early morning to resume the search for our rifles. No luck again for another hour. We were serious cadets and indeed worried that the D S could be coming soon, which meant serious trouble.

Another GC looking for his rifle yelled in excitement, "*Look,* come here quick, our rifles". We ran in the slush and rain and lo and behold, what do we see?

A fat overweight techy dope - GC Bajpai was sleeping calmly in pouring rain on a bridge of rifles made over a narrow drain in the field. He had collected all the rifles he could lay his hands on, put them across a small drain in the field to make a small bridge, spread some hay and a military rain coat and made a bed. Not too comfortable, fairly effective though. He slept peacefully covering his face with helmet, and his body with another 'whacked' water proof. All through the night, the rain poured with water gushing under him as techy Bajpai slept with gay abandon. And we, half dead, worrying about the consequences of not finding our rifles.

What was left to be done, techy dope Bajpai was kicked awake and roughed up nice and proper. He was served a multi course breakfast of choicest abuses and given a bath in the trench, where we had spent hours looking for our rifles. We were then worried about our rifles, concerned that the barrels may have got bent under his weight. Luckily, in the army they make the weapons soldier proof and all was well.

GC Bajpai was a civil engineer, an architect. He had put his knowledge to good use and made himself quite comfortable under the circumstances, never mind our plight.

IMA CAMP- TECHY'S INNOVATION;
WATER WATER EVERYWHERE NOT AN INCH TO SLEEP

The Incredible Techy

A techy was generally associated with doping. Coming from cool comforts and care of home, he found himself in hostile territory, where anyone who was someone bayed for his sweat, if not blood. Coming after studying laws of science, it took him some time to find his bearings. Everything in IMA was an unpleasant experience for them. Based on their own interpretation of rules some land up with a wrong notion of already having become officers, for them, the fall my countrymen, is even worse.

GC Munny, a techy dope who had failed in almost all the tests was called for a periodic interview by the Platoon Commander. The Platoon Commander read out his performance report and warned him to improve, failing which, he could be relegated. The instructor decided to give him a written warning.

"Read the warning and put your 'dhobi mark' at the

bottom." Having said that, the Platoon Commander got busy with another cadet's dossier. The techy read the warning, took a pen and scribbled on the paper, saluted and marched off. The next GC marched in for the interview.

When the Platoon Commander looked at Munny's dossier to see if he had signed, he was surprised to see a funny scribble which looked like this --**--. Funny signatures he thought. Call GC Munny, he snapped. The cadet bolted off and brought GC Munny back.

The Platoon Commander then asked him, "I thought I had told you to put your signature, what have you done?" The techy who by now was literally shivering replied sheepishly, "Sir, you told me to put my dhobi mark and I did that." (Dhobi mark is a slang in the army for signatures). The Platoon Commander smiled, just about controlling himself from bursting into laughter. I am sure he would have given full marks to the techy who amazingly remembered his dhobi mark of all the things. Incredible! Trust a techy to do the impossible.

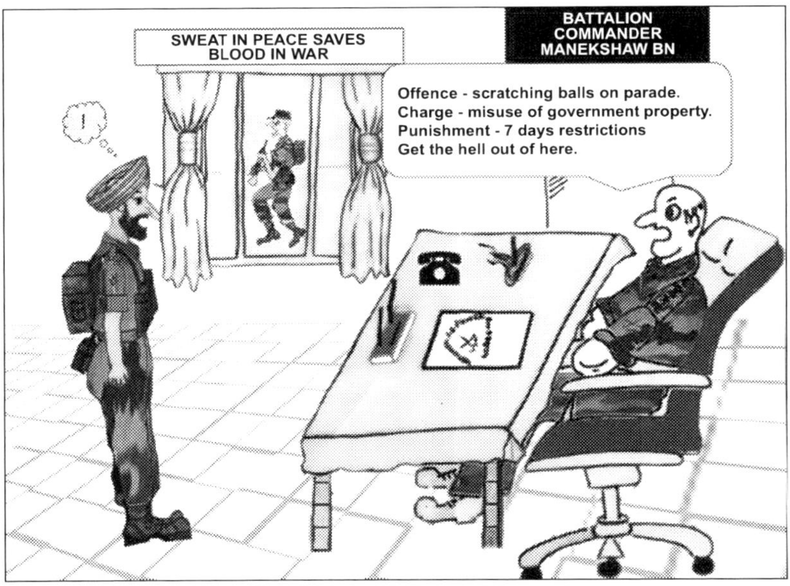

Do You Know Where You Stand? Yes I Do, Says the Techy

Some techies took too much for granted. Now that they were in the academy, they were the sons-in-law of the army, and it was their headache to see that they passed out from IMA. Well, it was not a wrong assessment. Army is a tough profession and there are not many takers for this wonderful profession amongst technical graduates in the country. So, truly, IMA went all out to ensure these techies did pass out, in the process compromising on many counts as far as the techies are concerned. They therefore tended to take things rather lightly.

During an outdoor training exercise on patrolling called Tactical Exercise Without Troops (TEWT), the instructor noticed that a techy, GC Khare was not interested in the TEWT. The TEWT consisted of a full day's discussion followed by issue of orders by the nominated cadet, followed by actual conduct of the exercise at night. To keep everyone on their toes, all cadets were asked to prepare patrol briefing. The instructor would nominate the patrol leader a little before the briefing.

The instructor had decided to nail GC Khare. The techy who thought he had nothing to do with patrolling was taken by surprise when he was nominated. He got up reluctantly, stretched his dungaree, dropped his cap on his head and moved to the cloth model made on a ground sheet. The cloth model itself would have earned the first prize for the worst effort. It contained bare essentials that included own position, a few irrelevant landmarks and enemy position.

Khare in full josh asked the platoon to sit at attention and took permission to commence the briefing.

"Aaj raat ko hamara section patrolling ke lie jaega. Kya ap

log patrolling jane ke liye tayar hain?" (To night our section will proceed on patrol. Are you all ready to proceed on patrolling to night?) The cadets who were always looking for fun shouted *"taiyar hain shrimaan"*. (Yes, more than ready, as if there is democracy in the army).

Task : *Hamara patrol inh sawalon ke jawab laega, kya ilaka jhari men dushman hai ya nahin, yadi hai toh uski tadad kya hai aur uska failao kya hai?* (Our patrol will bring answers to the following questions. Is there enemy in Area Bush, if so what is his strength and what is his extent?)

Route Out : "*Is prakar,*" (Like this) and moves his pointer in a sharp circular fashion from own position to the objective.

Route In : "*Is prakar,*" (Like this) and moves his pointer again in a sharp circular fashion from the objective back to own position.

"Kisi ko koi shak?" (Any one any questions?) and without waiting for a reply, *"Shak nahin toh ghari milayen. Jab mein bolunga time toh time hoga teen baj ke dus minute. Dus seconds.... Paanch, chaar, teen, dow, ek- Time- teen baj ker dus minute."*

(Any one any questions? If no questions, synchronize your watches. When I say time, it will be ten minutes past three.Ten seconds, five, four, three, two, one – Time- ten minutes past three).

The briefing had finished even before the section could settle down. The instructor who all this while sat quietly, apparently amused, said,

"My dear Khare, with your kind of briefing, no patrol will ever leave the unit, and if it does, the mission is a guaranteed failure. If you do manage to cross into the no man's land, half of your patrol is dead. The only prize you could expect to get with this effort is the prize for the worst patrol briefing ever, on either side of Suez Canal. You have been sleeping the whole bloody day. Now do you know where you stand?"

The instructor was not prepared for this one.

"Yes sir.." said Khare confidently pushing his chest out, "Here," pointing towards own position on the cloth model with his rickety make shift pointer.

Marya at his Innovative Best

Arun Marya was a sharp bloke. During one such TEWT he was taken by surprise and detailed to give the Ambush briefing. Like the others, he too was banking on his luck, which deserted him that day. No one had prepared the sketch trusting their luck, and worse, no one had even brought the ground sheet to make one, just in case.

Marya thought quickly and worked out something. His instructor, Captain Hooda, came along at the given time. He was impressed to see that all the cadets were sitting in proper discipline. Noticing that there was no sketch, he assumed, it was being prepared and sat down. Marya asked for permission to begin the briefing.

He then smartly took out a handkerchief and hung it from a branch of a tree. It had a sketch of the ambush site clumsily attempted. Marya completed his briefing with supreme confidence, swearing to make a mince meat of the enemy.

Captain Hooda's jaws fell at this comic scene. The class kept rolling for the next half an hour. He then appreciated Marya's improvisation and gave him full marks.

The Rattler Rattles Loud

During exercises in the academy a gizmo called 'rattler' is used to depict fire of a Light Machine Gun (LMG). It is made of wood and when rotated makes a repetitive *'terrrrr'* kind of sound.

During an exercise on platoon in defence, the instructor asked one techy,

"How will you deceive the enemy attacking your platoon?"

The techy replied, "Sir, when the enemy attacks I will place my LMG at one end and play the rattler from the other end."

The Light Machine Gun is not so Light After all

TEWTs such as the one above are pre – cursors and building blocks for training camps, which are the practical validation of all training. The second term camp involved a platoon in defence exercise. It was followed by a long – long endurance walk through the night and most of the next day, all that on a wet and freezing night and a scorching hot day.

The camp involved a lot of physical buggery. With almost twenty kilos on our back coupled with inclement weather, the endurance marches were a rather painful affair. To top it all, we were required to carry our personal weapons and a whole lot of ammunition, pick axes and shovels, ground sheet and many other essentials which made things even more difficult. A Light Machine Gun (LMG) was required to be carried one per section. Normally the LMG was carried by turn by all in the section. Anyone who has carried a LMG for long will tell you there is little truth in that statement. It is anything but light.

To add to the woes, GCs were required to carry a dozen and half-filled magazines with dummy rounds which is invariably the last straw on the GC's back.

Normally, the empty minds of GCs were never short of ideas to reduce the misery. Some platoons would strip the LMG in to five sub assemblies of barrel, butt, bipod, body and moving parts to share the load.

A good idea, but carried the risk of losing weapon parts which was an inexcusable blue murder. The LMG was ineffective as a weapon in this state and hence a wrong thing to do. If you were caught doing this, even God couldn't save your backside. So we decided against it and chose to carry it by turn in one piece.

It usually turned out that the cadet who was nominated to get the LMG issued on his name for the camp got stuck with it. He never found anyone around to pass it on during the route march. We thought that was unfair. It was decided that the LMG would be carried by turn for an hour each for the twenty four hours route march. The first cadet was selected by the draw of lots and then by turn based on the academy numbers. As the luck would have it, the ill luck fell on a small and physically weak techy, GC Goswami. He had always struggled to scrape through the physical training with great difficulty. One hour with LMG was manageable, he thought, and then it would be passed on. The first one hour walk was through the plains in day light. Thereafter, it was all through the high mountains at night that made him feel fortunate to be carrying it first. The gruelling route march started with friend Goswami bending dangerously forward to balance all the weight. Added to that misery was the LMG that seemed to be the last straw on his back.

For a moment I thought of taking the LMG from him. I discussed the idea with my buddy, GC Alan. "You must be a fool to even think of this. Don't be mad, do it when your turn comes. These buggers are also cadets and they must do their bit." Hungry, tired and half dead already by now, the endurance march commenced. The next eighteen hours over the mountains were a painful drudgery in cold and rain. After eighteen hours of screw of a walk, we were closing in with the academy, another three hours march away. By then we had been through with the hills and managing to drag our sore feet through the rocky bed of River Tons.

I was lazily ambling along wondering when the silly route march would finish. I noticed GC Sharma, an ex NDA, whiz past me at a rather frantic pace. Very unusual for that stage of the route march when walking itself was torture, I thought.

I asked him, *"Partner barhi jaldi mein hai, sab theek toh hai."* (Buddy you seem to be in a tearing hurry, hope all is well).

Sharma replied, *"Jara peeche mur ke dekh, sab theek nahi hai. Guru lamba kadam kar le peeche LMG aa rahi hai."* (Just look behind, all is not well. Partner step out, there is LMG coming behind). He replied without slowing down and disappeared out of sight with ever lengthening strides. The threat of LMG worked like a performance enhancing drug, giving Sharma inexplicable energy when everyone was struggling.

I turned around and was aghast to see friend Goswami, his big pack bouncing on his hips and his side pack and water bottle dangling at his knees. He was bending forward awkwardly and barely managed to keep the LMG on his shoulders. He was desperately trying to catch up with someone from his platoon to hand over the LMG. I later learnt he had been carrying the LMG right through for almost eighteen hours. Since the route march started, he never found anyone from the platoon to hand over the weapon. Smart as the cadets are, everyone made sure they were never within his reach. Once it got dark, and up in the mountains, finding someone when none wanted to be found, was simply impossible.

Armed with that timely warning, what was I supposed to do. I leave it for you to guess.

Captain Tura is Coming

The camps in IMA are a test of endurance, both mental and physical. The most sought after commodity is sleep and food. Both are in short supply. Cadets learn the art of sleeping with eyes open and walking with eyes closed depending upon what they were doing. The camps are often a battle of wits between the instructor and cadets, former denying the cadets sleep and the latter trying to catch some.

During the second term camp in the academy, Captain Tura, a hard task Platoon Commander told his company that he would like them to dig the trenches the whole night. Not a moment's sleep. Anyone caught sleeping, just watch your backside. The cadets are however incorrigible. At the earliest opportunity they would go to sleep and they did, leaving a sentry behind to provide early warning of the arrival of the instructor. At about mid night, the sentry in the trench, who was himself dozing off and on, noticed some one rapidly approaching in the dark night. He recognized the silhouette as Captain Tura. The sentry scrambled out of the trench and ran to wake up the cadets who were sleeping some distance away.

"Get up, get up quick," he whispered loudly "Captain Tura is coming, Captain Tura is coming." Suddenly he felt a firm hand on his shoulder and heard a voice, "Captain Tura is not coming, he has already come." The rest, as they say, is history.

G C Shukla, a DE, struggling throughout his stay in IMA had managed to fracture his leg during the 'Camp Chindit'. With a broken leg and a hand in sling from a fall, he bared his heart during the campfire programme of the camp. He sang this parody of a film song with real pain in the heart *"IMA ke GC re tera dard na jane koi...."*(Oh Gentleman Cadets of IMA, no one knows your pain...)

Shandy in a Suicidal Mode

Shandillya, known as Shandy, was the naughty bloke with the laughing toy in NDA. He was a perfect team man. He was always there to help everyone, whether in need or not, quite often at his own expense. He was particularly active with academic instructors.

We had appeared in a Military history test in the third term. When the papers were shown to the cadets, GC Ashok Pathak was marked poorly in a particular question. Sitting next to him, Shandy, always a volunteer for social service, read his answer and emphatically said that their answers were almost the same and he was given seven out of ten whereas Pathak was given only four. He egged Pathak to take on the instructor, this time from the Army Education Corps (AEC) , and demand more marks . He gave Pathak his paper to show the instructor. Pathak aggressively walked up to the instructor with Shandy's paper as documentary evidence in support of his demand. Vocal support from Shandy from the back bench was consistent.

The AEC Colonel paid no heed to Pathak's loud protest and refused to see Shandy's answer. He threatened disciplinary action if Pathak did not go back to his chair. Dejected, Pathak walked back to his seat slowly.

Shandy was the 'Last man - last round' variety. He was not going to give up so easily. He stood up and canvassed Pathak's case. He emphatically said both the answers were almost similar, yet there was so much disparity in the marks, four for Pathak and seven for him. The instructor asked Shandy to shut up and sit down, but he would not relent and kept on. The Colonel then said,

"You say both of you have written the same answer and he has been given four and you have been given seven."

"Yes, sir" replied Shandy with hope in his voice.

"OK, bring both papers here," said the instructor.

When he returned, he had a bewildered look on his face and a mischievous smile. The Colonel had reduced his marks from seven to four.

That was military justice. They were now at par and none could complain. Shandy had a good laugh at himself after the class.

John T's Parting Advice

As the final days of our training at IMA approached, the final rehearsals of the passing out parade commenced with the Adjutant's parade. His words of advice communicated through a story as he addressed the cadets from the podium.

There was a poor man. He had a large family to feed, some sheep and a cow. Life was hard and all that he had to live in was a small thatched hut. He decided to pray to Lord Jesus to help him out. He went to the top of the hill and prayed and Jesus appeared. He complained of his difficult life and prayed that Jesus give him a bigger house where he could live more comfortably.

"How many members in the family?" Jesus asked.

"Six children," the shepherd replied.

"How many animals?" Jesus asked.

"Five sheep and two cows, my Lord," the shepherd replied.

"How big is your room?"

"Very small, no more than ten feet by eight."

"OK, go back to your hut," said Jesus.

"But I requested for a bigger room, Lord Jesus."

"Yes, go back and take all your cattle and sheep inside the room and start living with them."

"How can I? There is hardly any space with such a large family. It is simply impossible, my Lord." But then he had to bow to Jesus' desire. The shepherd went back and took all his animals in. Life became terrible with his children having to sleep under the animals with all the filth that they created. He cursed Jesus.

Somehow, the shepherd passed his seven days in misery and returned to meet Jesus.

"How has it been?" asked Jesus. The shepherd listed his catalogue of miseries with all the animals inside his small hut and requested Jesus for a solution.

"Alright," Jesus said, "Go back. Take all the animals out of the hut when you sleep tonight and see me tomorrow."

The shepherd went home, removed all his animals out and slept without them. He went back to Jesus who asked him, how was it? The shepherd replied, "Oh, Lord Jesus, it is simply wonderful, there is so much space for all of us." John T's message, live a simple life with controlled desire. He himself was a simple man with simple life.

Another thing he mentioned was the issue of seniority that we get stuck with, either when we meet personally or on phone, both reluctant to sir each other. His advice, your seniority does not change even if you sir a junior. Soon the relative seniority will be known and the issue settled. The junior will only be more respectful for your largesse.

The Final Step

Like the NDA, GCs pass out of IMA to the tune of 'Auld Lang Syne' watched by their proud parents. One small step over a marble slab with the inscription 'The Last Step' (leading into Chetwode Hall) catapults them into a completely different league. He turns into a confident young officer entrusted by the President's Commission to safeguard the honour, dignity and integrity of the nation. At the young age of twenty one, he would be entrusted with the lives of over a hundred men. They would follow their young officers even to their death, for these young warriors would lead from the front, sharing every hardship and risk with the wonderful men they command.

If you haven't visited the IMA war memorial, I suggest you do so. The long list of martyrs will speak loud of their service to the nation. Remember Kargil war of 1999?

PAHLA HAMESHA PAHLA

Role of Brigade of The Guards

"To take its proper place in War and acquit itself in an exemplary manner in action and discipline".

Motto

Pahla - Hamesha - Pahla

PART IV

THE REGIMENTAL REMINISCENCES

The transformation from the highs of passing out from IMA to becoming the junior most officer in the battalion is a mixed bag. From being the Adjutant or the Academy Under Officer with all the authority at your command, you are reduced to almost nobody. On the other side, a young officer's arrival brings a lot of excitement and hope in the battalion; hope over the possibility of getting an excellent youngster. He goes to the battalion with a lot of anxiety and nervousness, wondering what lies in store. The reception that a young officer receives on arrival varies from regiment to regiment and from unit to unit. Some may find themselves being brought in the tank of a water truck from the railway station. He begins from ground zero, starting from learning regimental history and gets more and more confident as days go by. Slowly he settles down and his battalion becomes his home away from home. It then becomes part of his soul, an association that lasts beyond his life time and extends to his children who grow up in the battalion. His twenty years in the battalion will be the best part of his life and he can never forget those.

The armies all over the world are similar in almost everything they do. I have had the opportunity to see some of them at work. Things that make them similar are the common values of discipline, esprit de corps, camaraderie, common nature of risks and hardships as part of the job, pride they take in being soldiers and accomplishing the mission assigned - no matter the cost, purely for professional pride and unit's izzat.

Well, Indian Army is a bit different and very special. Very special because of the regimental system which is the strength of our army. A legacy of the colonial days, the regimental system is prevalent in only few countries, conspicuous being India, Pakistan and of course UK. The Indian system remains special because of the cultural, religious and regional diversity of the country. With over two dozen Infantry regiments, each with a different character, customs and traditions, it is a great institution. These regiments are quite distinct from what the western armies understand, where a regiment is a formation like a brigade with three battalions. In India, Infantry is composed of over two dozen infantry regiments, representing different regions of the country. For example there is Punjab regiment based on the troops hailing from states of Punjab and Himachal Pradesh. There is Bihar Regiment, with its troops from the state of Bihar and around, then there is the Sikh and Sikh Light Infantry regiments which has only Sikh troops. Then there are number of Gurkha regiments which have only Gurkha soldiers. Like this there are many such regiments, each with over a dozen battalions. While all the regiments wear OGs as the basic uniform, they have differently coloured regimental accoutrements that gives them a distinct identity. All battalions of a regiment wear the same uniform, draw the men from the same region, have a common training centre and are part of one large family. They have one 'Colonel of the Regiment' who looks after the affairs of the regiment. There will be scores of men in a battalion from the same village, many related to each other. The second and third generations will join the same battalion. There could be over a dozen father and son or brothers serving in the same battalion at the same time. There could be three successive generation officers commanding the same battalion in their turn. The officers and men serve in the unit they join for the entire duration of their regimental service which is nearly two decades. Officers mostly command the same unit. So the camaraderie is supreme and officers and men know each other better than their mothers do, as they say in the army. Here, as we say, the officers are naked. Eight hundred pairs of eyes watch you every day, for two decades. There is a typical military way of putting it across but let me put it more politely. The whole battalion knows what colour underwear the officer wears or does not wear one at all. So there is nothing unknown, the men know your true colour under the skin. The izzat and the honour of the battalion guide everything that its officers and men do.

It is not uncommon to see a senior JCO or NCO walk up to a newly posted Lieutenant, telling him that he has played in his arms as a toddler, when he was sahayak to his father. The officers' kids in the battalion would grow up together and then many of them would be officers in the same battalion. When I was commanding my battalion we had five of them, whom I had seen as toddlers when I had joined as a young officer on commission, and that is absolutely wonderful. An officer's son could be calling every one uncle one day and a few months later, would be there in uniform clicking heels and saying 'yes sir.'

Traditions have their own place in the army and they are impossible to change because of sentimental values. Many of them make no sense at all, but they are there to stay. 3 Kumaon Rifles holds it regimental cane the wrong way. That is the knob end facing backwards, where as it is meant to be held the other way. Ask an officer of the unit, he will tell you that is the right way. How did that happen? I am told that a senior officer was visiting this unit. The officers had lined up for introduction to the VIP. The commanding officer receiving the visitor held the cane the wrong way, that is the knob backwards. The second-in-command noticed that alright, but too late to make a difference. He then did, which only the army can do. He instructed other officers to reverse their canes and hold it the way the commanding officer was holding. It was done, no one noticed the fault and the battalion has since then been holding their canes the wrong way, their right way.

The Paratroopers have their own way of life. Many years ago, there were two officers from the Parachute Regiment. Let us call them Jakes and Garry. Jakes, a para commando, was a brilliant pistol shot and had been participating in national shooting championship. Sitting in a bar during a para refresher course in Agra, they were drinking hard, the way paratroopers did those days.

Well past midnight, Jakes said he will demonstrate his shooting skills and asked Garry to stand against the wall. Garry stood against the wall without blinking an eyelid. Jakes loaded his pistol and from the other end of the room fired a number of times stringing a ring of shots around Garry's head. Impressed by Garry's trust in another paratrooper, Jakes said he would reciprocate the trust. Jakes did not blink as he handed his pistol to Garry and stood against the wall himself.

With more than a bottle of rum between them, and Garry having no claim to shooting fame, one paratrooper's blind trust in the other prompted Jakes to risk his life.

The tipsy trooper stretched his unsteady hand and fired. Thanks to his poor shooting and rum, he missed the huge wall. The bullet went through Jakes' cheek and neck as he dropped on the floor. In a flash, the fun had turned into a disaster of the most serious kind. Certain that he will kick the bucket, Jakes realized the implications for Garry that could land him in jail for life. He asked Garry to get a pen and paper and wrote a suicide note. He asked Garry to wipe his finger prints from the pistol. A medico legal case, no doctor would touch him with a barge pole. Eventually the incident was successfully managed and

Jakes lived to shoot another day. As for the civilian barman, he took flight the moment the first shot was fired. He was retrieved from his hiding after a week.

So much for the paratroopers' incredible camaraderie. It is always risky to invite a paratrooper to a drink. He will drink your whisky and if he is in a good regimental mood, eat away your whisky glass too.

Interestingly, each of the regiment has its own character, and we understand that. I remember a divisional commander mentioning, "Go visit a Jat battalion, and they will serve you sweet glucose biscuits with soup against the backdrop of mess curtains in regimental blue." An officer from Assam regiment starts talking like their troops with the same accent, clad in jeans and T shirts the way their troops do. Some from the Sikh and the Sikh Light Infantry will have an unusual sense of dressing and of course their flowery language and so on. In fact, one can often guess the regiments the officers belong to with a fair degree of accuracy.

Being mostly in difficult operational environment, infantry officers are battle hardened. They carry a vast variety of experience no one else does. There are other arms and services that have their own kind of experience and may not have crossed the line joining Ambala and Patiala.

Our men coming from all corners of the country with different religions and cultures, speaking different languages are a wonderful lot. Indian Army's history is a saga of valour and sacrifice. They have achieved almost impossible feats in battle. They can smile and laugh in most difficult conditions.

In this part, I have narrated fascinating experiences from my time in my wonderful battalion-5 GUARDS where I served for twenty two years.

Regimental Reminiscences

I was commissioned in 5 GUARDS in June 1979. That was previous century, feels like a dinosaur, except that I am not extinct yet. Well in IMA, I was an optee for GUARDS missile battalion for some reason. Despite being on top of the heap, I was allotted 5 GUARDS, an infantry battalion. Then Captain Lidder from Grenadiers, an instructor at IMA asked me.

"Believe you are going to 5 GUARDS, but I know you had opted for GUARDS missile battalion."

" Yes sir," I replied.

"So are you asking for a change? You are entitled to get what you asked for. I do not know the missile battalions, but I have served with 5 GUARDS. All that I can tell you is, if you go to 5 GUARDS, you will never regret," he said.

I said, "Sir, you are from Grenadiers, you have nothing to take in the matter. But if you think that way about 5 GUARDS, that is the way it stays and I am going to 5 GUARDS then."

I joined the battalion and it was the wisest decision I made. It has been a story of two decades of wonderful soldiering, for 5 GUARDS was forever in field and operations and was rightly known as 'The Fighting Fifth'. Most of these years were spent in most difficult of places in difficult environment. It was precisely these difficult times that brought out the best from these wonderful highlanders hailing from the lengths and breadths of Himalayas from J&K to North-East. These wonderful soldiers never complained of anything and brought the battalion glory, time and again. These soldiers gave us great moments of pleasure through their simplicity, dedication, loyalty and above all their sense of humour.

It is the wonderful memories from that period that I wish to share with you here.

The Bedding In: Baptism by Fire

A battalion is a great institution; a home away from home. It is the strength of our army. Good battalions- good army, bad battalions- bad army. It is called a unit, which means (1) one, because despite over eight hundred men, it functions like one entity powered by common ethos and culture, soul and spirit and one mission and hence a unit.

My battalion has a very fine class composition with troops hailing from trans- Himalayas, starting from Jammu and Kashmir, Himachal, Kumaon, Garhwal, Gurkhas and the seven states of the North-East. These simple souls were wonderful soldiers that made our life in unit an absolute pleasure. They were disciplined, enthusiastic, well motivated, physically fit, well turned out and above all, most loyal. Three decades on, I do not remember one instance when they have been found wanting. I have had the honour of being with them in peace and operations. They were superb and simply never complained hell or fire, blistering heat or freezing cold, thirsty or hungry. Our missile platoon had participated in Kargil war and won two Sena Medals. They fought for over two weeks simply on water and biscuits. In the debriefing held at the unit, they never even mentioned it till we asked. The time spent in the battalion is simply unforgettable. It has never been the same since leaving the battalion on relinquishing command of this wonderful lot.

There are many memorable experiences that I have narrated in this part of the book. The life those days was different. The cross border terrorism in Kashmir had not started. There was ample time for unit routine, PT, games, training et al. Inter battalion competitions were contested with great zeal bordering on conflict at times. Young officers spent most of their time learning the tricks of their trade, participating in inter battalion games, learning regimental history etc. Thankfully there was no cable TV those days.

I had my first unique experience as Mr. Sengar, as the Second Lieutenants were known those days, during the Dussehra celebration in the paltan in 1979. Being a hill tribes battalion, Dussehra was celebrated in a big way over two days. Past midnight, each company would sacrifice one goat that was served as prasad with a tot of rum with the whole battalion present.

My commanding officer (CO), Lieutenant Colonel Pup Mann was an excellent flamboyant commander. He would make it a point to project his officers wherever opportunity presented. As the sacrifice ceremony was on and the goats were being sent to their heavenly abode, one at a time, the CO directed that I do the honour and slaughter one. Never to say no, I was there with a khukri in my hand. The Subedar Major walked up to me and cautioned me to be careful and not to hurt my legs in the process. That was a good advice as I had never handled a khukri before, let alone slaughter a goat. His caution made me rather cautious. The goat was dragged to the altar, after kind of forcing its acceptance to be sacrificed by pouring half a bucket of cold water. I stood there with a white cloth tied around my head like a samurai warrior. As I swung the khukri through the goat's neck, the goat struggled and moved. The khukri slashed through its neck, leaving a sliver of flesh connecting the head to the body as it slumped to the ground. I struck again to complete the job.

Well, I had failed in the task to chop off the neck in one go; never mind the margin of failure. I was penalized by the CO - a peg of Hercules rum to the whole battalion. It cost me better part of my month's pay. Great memories. Three decades later, I can recount the whole evening to the last detail.

Another occasion, this time in Sevok Road, Siliguri 1979. We were having a Sunday family get together in the officers' mess. Pineapples were common and that was the fruit on the menu. As the last one to be served, I got a small piece of pineapple to eat. Pup Mann asked me how many pineapples I could eat in a sitting. Small piece that I ate had felt good leaving me wanting more. Without much thought, I said, "eight". Eight pineapples in a sitting! No one believed that I could. As for me, I did not realize what eight pineapples meant. Tuffy Marwah, the spirited second- in - command laid a bet that I could not. With a number of beer bottles at stake, the stage was set.

The CO gave his directions and the next day eight pineapples were displayed outside the second-in- command's office. All officers, some JCOs and NCOs and mess staff formed a fairly impressive spectators gallery. I wasn't impressed to see that all the eight pineapples on parade, lined neatly in rows of two were rather raw, poorly peeled and possibly the largest available on either side of Brahmaputra. Tuffy Marwah had been smart. There was no way he was going to lose the bet. Even though I protested, a second Lieutenant's protest had no future.

I was not giving up either and started eating the pineapples. With great difficulty I managed to finish one. By the time I got there, my mouth and teeth had become lifeless unable to even bite the slice of the fruit. The citric acid and thorns from the raw and poorly peeled fruits had cut through my tongue and taken a toll. Blood was dripping from my mouth and over my chin. As I continued to battle valiantly, the CO asked me to stop.

He said, the experience will make me a better man and I would be more deliberate in task analysis in future. Interestingly, I offered the officers to join in the feast. None did.

Soldierly Logic

Lance Naik Ganesh Bahadur Gurung, a happy go lucky Gurkha, was my Sahayak (termed batman those days) when I joined the unit on commission in 1979 in Siliguri. The weather in summers was far from good. Adding to that was the terrible mosquito menace. In the good old days, there was no luxury of generators, as it is now. So your best bet against the mosquitoes was, as it is today I guess, the ever reliable mosquito net. These days there are multiple options to keep the mosquitoes at bay, then there was just one.

One night was a particularly horrible night. The mosquitoes had me nice and proper. Looking at the mirror, I saw my face loaded with mosquito bites. I summoned friend Ganesh Bahadur and pulled him up for not tucking the mosquito net properly and showed him my face. Coolly he smiled and asked me, "Saab, do you know what the purpose of the mosquito net is?" I said, I think I know. Before I said anything further, he continued, "The mosquito net is meant to trap the mosquitoes, so that they can be killed and do not bother again the next time. So now let me keep the mosquitoes inside and after you have gone to the office, I will kill them peacefully. Just see how well you sleep tomorrow."

Well, I knew it was pointless to argue with him and his logic, and I knew the next night would be no different and it was not. Hail Ganesh Bahadur Gurung.

Faux Pas Most Royale

It had been barely three months of my commission that my Commanding Officer (CO), Lieutenant Colonel HRS Mann, popularly known as Pup in the regiment, entrusted me with my first important assignment. The Indian Army was then carrying out trials for introduction of the disruptive camouflage combat uniforms, called combat dress. Mine was one of the battalions earmarked to carry out the trials in our sector.

There were three types of uniforms with different types of disruptive pattern. The uniforms had been in use for almost six months and were beginning to show their true colours. The Army Commander was to visit the formation shortly and one of the many items on the itinerary of the General was a briefing on the trial of the combat uniform. That is one big occasion for a brigade. Army commanders don't visit brigade headquarters every day. Success or failure of these visits could make or mar many a careers. My battalion was tasked for that briefing to the VVIP. In keeping with his style, Colonel Pup Mann detailed me to brief the Army Commander. The Brigade Commander apparently was not happy seeing a young 'recruit' given the important responsibility. He would take no chances and said he would like the battalion to carry out rehearsals and get the whole thing approved from him. The brigade headquarters was selected as the venue for the briefing.

Number of rehearsals with the CO in attendance were carried out till the commander was finally satisfied. The final format approved was that three soldiers would stand wearing six months old uniforms of different patterns and three brand new sets would be displayed on a table for the purpose of comparison. In conclusion I was to give out the pattern recommended for the army. On the last rehearsal, the day before the arrival of the Army Commander, Pup Mann told me that he was not satisfied with the 'monkeys,' as he used to call the troops affectionately, selected for the briefing. Change them. Get smarter guys, all of one complexion either dark or fair. They were all smart I thought, all six footer Guardsmen, except that two of them were dark complexioned Assamese and one was a fair Gurkha. 'Order is order and there is not to question why'. Only, I did not realize at that stage how close I was to my professional grave, which the whole episode would lead me to.

Well! I got back, got hold of the Battalion Sergeant Major Kul Bahadur and told him that I wanted three fair NCOs for the job. No problem *"Hoon chha huzur"* (will be done saab, he said confidently).

The next day we waited in our officers' mess for the arrival of the Army Commander where the introduction of officers was to take place. I was then to move to the brigade headquarters for briefing the General on the combat uniform trial. Col Pup Mann asked me "AK are you all set?" "Yes sir." I replied. "Are you sure?" he asked again. "Yes sir," I replied again.

"I am not so sure," he said. "I want you to go and see things for yourself again and get back here in time."

I took off in a rickety One Ton vehicle whose door had to be held to prevent it from falling off. When I got to the battalion, I realized how correct the CO was. Since the three soldiers were changed the last day, there was utter chaos. None of them was around. I got hold of the Battalion Sergeant Major who ran around to get these fellows. They trickled in one by one. As they came in, I wasn't satisfied with their turnout. They were not briefed so they had no idea what they were required to do. I sent them to either polish their boots or brasso the belt or something else, as a result, I never saw all of them together. Finally, seeing that the time was running out, I told the Sergeant Major to get these guys and send them to the brigade headquarters at the earliest. I rushed back to the mess where the officers were waiting for the Army Commander.

"All ok?" asked Pup. "Yes sir," I said confidently.

The Army Commander drove in led by the military police pilots. The officers were introduced and the General proceeded to the next unit. I took off for the brigade headquarters to take stock of my mission. Like a good soldier, I decided to carry out a last minute check. I lined up the three NCOs. The earth under me slipped. Of the three NCOs, two were wearing the uniform with the same disruptive pattern. The one that was missing, was the one we had recommended for retention. I guess, Second Lieutenants, out of sheer ignorance of the gravity of situation are better geared to withstand a crisis like this. It did not perturb me much. I walked up to the camp commandant, a young Captain and told him the problem. I asked him for a vehicle so that I could send the NCO to change the combat dress. The camp commandant, not much senior himself, didn't think much of my problem. I guess my problem was my problem, he had his own problems with the Army Commander expected to arrive shortly. He did nothing about it.

8 JAT had recently arrived in the station. Their officers were to be introduced to the General in the brigade headquarters. About a dozen Jat officers had lined up a little distance away from where I was to give my land mark briefing that would change the way Indian Army would dress up. The vehicle I asked for was nowhere in sight. When I heard the blazing sirens of the General's motorcade, there was nothing that could be done any more. I decided to break off one of the two NCOs wearing the same pattern of uniform and told him to vanish out of sight. I lined up the other two and waited coolly for the arrival of the Army Commander. The Brigade Commander who had come out of his office on hearing the siren, looked at me and asked,

"Are you all set?"

"Yes sir," I replied.

The Commander sensing that something was amiss walked

up to me and asked, "Where is the third soldier?"

For want of a suitable reply I stupidly shot off what came to my mind, "Sir, I have broken him off."

Unable to believe that, he said, "Broken him off ! Why the hell?"

"I thought three were far too many." I gave the silly reply, not quite knowing what to say , getting myself deeper in the hole.

Not used to subalterns giving silly replies, he said, "Far too many, you think I am a fool wasting my time for the last seven days seeing the rehearsals here." The blazing sirens could be heard louder. You could now see the motorcade speeding towards the headquarters not too far away on the straight national highway. I stood quiet. There was little else I could do.

Observing that the pattern that was recommended was not to be seen, he asked, "Where is the pattern that we have recommended?"

"That is the one I have broken off sir." That was the coup de grace, the last nail in my coffin. I could imagine what was going on in his mind. If he could have his way, he would have buried me alive.

Seeing the hopelessness of the situation, he cooled down, like I was. "What can you do now?" he asked.

"Sir, I can make one of them wear the new pair of the recommended pattern displayed on the table, and draw comparison with the old one that the second soldier is wearing."

"Do that then," he said.

"But I don't have the cap for it." I replied (It was not part of the trial dress). He had given up by this time. I am sure by now he had written the obituary of his career and mine and possibly my commanding officer's. The motorcade rolled into the headquarters. As the second -in - command of 8 JAT commanded his officers to attention, the Commander sprinted away to receive the General.

Quickly I instructed one of the NCOs to put on the new

pair of the recommended pattern. As the Army Commander started meeting the officers of 8 JAT one by one, less than twenty yards from me, my Assamese NCO in his blazing multi coloured underwear was struggling to get into the new uniform. The buttons and button holes of newly stitched uniforms are always difficult to manage. It was a race against time. My jawan somehow beat the clock and managed to squeeze himself inside the rather ill fitting uniform, balanced the crooked ordnance factory cap on his nut and stood like a statue. As the General shook hands with the last officer of 8 JAT and turned, my team of three musketeers was just about ready for the show.

The General's informal approach did not allow me to launch my prepared briefing. He simply said, "Irrespective of what you think about it young man, the decision has already been taken. From the Chief to the last soldier, everyone will wear the combat uniform shortly." In addition to the 'faux pas' galore I had created, the whole format of my briefing went for a six. I spoke not a word, only heard. Only the Army Commander spoke and none of our recommendations saw the light of the day.

I had no doubts in my mind that it was 'one big bugger's muddle'. My first mission as a commissioned officer was an unmitigated disaster. If there was a provision, the Brigade Commander would have summarily reduced me to next junior rank. Since the next junior rank was a Subedar Major, who is a JCO, I survived to live another day. As I said, the subalterns by virtue of ignorance are better geared to weather the storms. It did not bother me much, since at the end, it did not seem to have created an embarrassment for the Commander. For that reason I did not tell anyone in the battalion what had happened and forgot about it.

Not many days later, the CO, Colonel Pup Mann walked into the mess at dinner time. "How are you AK?" he asked.

"Very well, thank you and yourself, sir," I enquired.

"I am good, thank you," he replied.

"Here is a present for you," and taking out a used cravat from his pocket handed it over to me. I accepted it with pleasure thinking it must be for some good job done. After all, commanding officers don't present cravats to a subaltern for nothing. After a drink he left the mess. Major Tuffy Marwah, the second- in - command who was a bachelor and witness to the event in the mess called me to his office the next day and said:

"Do you know why the CO gave you the scarf?"

"Not really sir," I replied, "I guess he must be pleased with me for something, why else would he do that, sir."

"Like hell. It is for the grand goof up you created on the briefing for the Army Commander. The CO got a letter from the Brigade Commander expressing his displeasure, the only one in his command. He thinks that you would remember longer with a present than you would with a reprimand."

He was right. Thirty years later, I remember it and always will. I learnt many lessons. One, on the day of the incident, and the second - the bigger one - on the day I learnt what the scarf was for. I guess those days are gone. Today's is a zero error syndrome army. Not many commanding officers would take it that way today, I am sure.

The episode would be a waste unless I list out the lessons from which others could benefit as well.

- On important occasion, never make major last minute changes unless inescapable.
- Don't leave important things to NCOs, as I did.
- Nothing checked is nothing done.
- Make a check list. Carry out inspection when there is still time to address the faults.
- Keep the spare resources handy. In this case, I should not have broken off the third soldier. If he was available, he could have worn the third uniform and stood there embarrassing the commander a little lesser.
- Don't give stupid replies as I did. Accept your mistakes. The

truth will set you free and the agony will last much shorter.

In this serious business, there were funny moments alright. When the Army Commander was meeting the last of the Jat officers, my happy go lucky jawan, in his multi coloured underwear was struggling with the pant buttons of the new uniform. When he finally did manage that and put the cap on his head, true to the Ordnance factory reputation, it was the most crooked one I have ever seen. As they say, 'When troubles come, they don't come in singles. They come in battalions and come at a time when they are most unwelcome.

Navigation by Stars

A senior subaltern in my unit had the habit of asking offhand questions to ensure that the youngsters kept themselves updated. He had earlier taught them to find the cardinal directions from stars. One night, returning from dinner on a clear sky with twinkling stars, he asked one Second Lieutenant, "Which way is the North, Ravi?"

After rotating 360 degrees, gazing hard at the sky for a while, he said, "That way sir," confidently pointing towards what actually was the South.

"How do you know?" the senior subaltern asked. "I saw it during the day time," replied Ravi.

Congrats Mate

A letter from Commanding Officer, 5 GUARDS, to a battalion officer on being blessed with a daughter. The husband in this case was not a particularly handsome man; short, deep seated eyes, high cheek bones and sharp features et al. His wife, on the other hand was a beautiful charming lady.

"Heartiest congratulations on having been blessed with a daughter. The Paltan wishes her good health and happiness. Further, if the new pocket edition is half as pretty as your wife, you would have done a great service to the humanity. God bless."

Beware of Free Advice

Colonel SR Singh was a humorous commanding officer (CO). Sometimes his sense of humour hammered others. Located in Meerut, he invited all brigade and unit officers for a Sunday brunch with their families and children.

As the kids ran havoc in the mess garden, the CO of a neighbouring unit remarked "SR, I notice that your officers have mostly boys and hardly any girls. Whereas in my unit it is the other way round. Any secrets?"

"Yes, indeed there is a secret," replied Colonel SR.

The other CO who had one daughter suddenly seemed interested in the discussion. Possibly he was planning another kid, and like other Indians, was hoping it would be a boy.

"What is the secret yaar?"

"Well, the secret is..." SR started slowly, generating suspense as his interlocutor looked on. "The secret is that on the night that you plan to have a baby, the entire ambience should be pinkish. As you enter the room, it should give you a soothing pink feeling. To achieve that, have nice crisp pink bed sheets, pink pillow covers, pink mosquito net and pink night bulb. You should wear a pink night suit. Similarly, the lady should wear a pink nighty, and don't mind, the lady should also wear pink undergarments. Keep two pink pairs of slippers. And remember..." SR paused and gazed at wilderness as if thinking deeply. His friend looked at him in anticipation, waiting for him to continue.

"And yes, you need to choose a full moon night for the all important event. Light a lamp of pure ghee in sight of the moon. Ask the lady to climb the bed from the northern end and lie down on the left side of the bed ..." Colonel SR paused for a moment as his friend looked on eagerly.

"And,... and when everything is ready,............ call me."

Well..well, his friend could not have been amused for sure.

Dal Bahadur and his Logic

Some johnnies can be trusted to do things their way, no matter what the orders are. My battalion had a large percentage of Gurkha troops when I joined the unit in 1979. The battalion was posted in Sikkim. It was a high altitude area with pretty dangerous roads. There were strict instructions that vehicles halting on the road will place a wooden block, called 'Gutka', in front of the wheel to prevent the possibility of vehicles rolling down.

One day, as a subaltern, I was delighted to be given a one x 1 ton military truck to run official errands. Mostly it was the rickety 3 Ton Shaktiman which made you feel you were perched atop a two storey building. My chauffeur was Havaldar Dal Bahadur, a pleasant natured Gurkha bloke. He stopped the vehicle next to a Military Post (TCP), got down, placed the block behind the wheel and proceeded to report to the TCP.

Since he was in a queue and it was taking him time, I got down and happened to look at the block. To my amusement, I saw that the block was placed on the wrong side of the slope. The vehicle was on a down slope whereas the Gutka was placed behind the wheel. When I asked him, he simply said, *"Saab fikar nahi, gari gear mein hai, block toh CMP (Corps of Military Police) ke dekhne ke liye lagaya hai, nahin toh report bhejta hai ye kambakhat."* (Saab don't worry, the vehicle is in gear and the hand brake is on. This block is only for CMP to see, otherwise these clots send an offence report to the unit).

In those days of archaic 1 Ton and 3 Ton vehicles, the handbrakes were mostly ceremonial and rarely, if ever, functional. Many accidents had taken place because the handbrake did not work and vehicle rolled down the hill.

He, however, learnt his lesson soon. At the next stop he did the same thing. The vehicle started moving down the slope as the gear slipped into neutral, leaving the block and

the ineffective hand brake in place. And when that happened, friend Dal Bahadur ran in front of the vehicle trying to stop it like a sumo wrestler. Fortunately, I happened to be in the vehicle and managed to slide my leg through the maze of gears and push the foot brake, averting a disaster. Thereon, it was a sight to watch friend Dal Bahadur, almost lying down on the road to ensure he had got the slope right before placing the Gutka.

Chhetri Gets out of Tight Corner

Some Gurkhas are real smart thinkers. In 1979 we had Subedar Bodhraj Chhetri, an outstanding Gurkha JCO. He was 'Delta' on Mortar Course and was the Mortar platoon JCO.

The Mortar Platoon is the best platoon in an Infantry battalion with the best jawans posted there. As usual the platoon was carrying out routine training at their squad post. More often than not, Mortar platoons continue to practice the same drill day in and day out, using the same data and facing the Mortars in the same general direction. Since I had done the Mortar course, I had a good idea about the whole thing. One day during a visit to the training area, I decided to change the whole setting to see how they fared. I painted a scenario of own advance from the opposite direction to the one they had been practicing every day. Subedar Chhetri was nominated the Mortar Fire Controller (MFC). I told them I would be noting the time taken by the platoon to deploy and bring down fire. (The MFC moves right up front with the leading companies. When required, the Mortar Platoon, moving quite some distance behind is deployed , and brings down fire directed by the MFC on radio).

The platoon stripped the Mortars and started advancing.

The advance of the battalion was halted by the enemy position. The MFC ordered the platoon to deploy and provide fire support for the battalion quick attack.

Frantic activity started as everyone barked out their orders and ran helter skelter, which would seem chaotic to those unfamiliar with the Mortar drills. Finally, the Mortar Platoon was deployed and 'Ready to fire' report was given. Subedar Chhetri, the MFC, looking at his watch told me with a wide grin that they have deployed in excellent time, clearly pleased with the performance and looking for a pat on the back.

I was not surprised to see that the Mortars were facing the same direction, in which they had been training every day, full one hundred and forty five degrees off the given direction of advance. When I was dressing them down on the poor show, Subedar Chhetri interrupted and said,

"Ekdam theek pakra saab aapne. Tub se mein dekh raha hoon ki yeh Mortar ek dum galat disha men deploy kar rahe hain." (You have perfectly caught the mistake, sir. Since long I have been watching that they are deploying the Mortars in the wrong direction).

"Aap toh Mortar Platoon JCO hai, toh aapne bola kyon nahi saab?" I said. (You are the Mortar Platoon JCO, why didn't you point out then?)

"Mein kaise bolega saab. Apne toh mujhe MFC banaya hai. Mein toh company ke saath aage hoon. Mujh ko thori dikhai de raha hai peeche yeh platoon kya kar raha hai." (How can I? I am the MFC with the companies ahead, I cannot see what the platoon is doing so far behind).

We laughed and appreciated his wit. I understood why he earned Delta grading on the course. Well deserved.

We Won't Let You Go Honey

- *Mess Havaldar- A selected Non Commissioned Officer (NCO), responsible for upkeep and management of Officers' Mess.*

Posted in Ladakh, in the late eighties, we all camped in tents. The officers' mess was a frugal snow tent. Join two of them and you had a reasonable place to eat in the comfort of a kerosene fed bukhari (Tin cylinder with kerosene oil burner. Military issue in extreme cold) that kept the tent warm. The whirring sound of bukhari was Beethoven classical music to an infantry officer's ears, sweeter than anything else in the world. The sound was an assurance of a warm night, except that a back blast from furious winds could convert that life saving bukhari to a life threatening one. The tent could go up in smoke before you could say "Jack Robinson".

Well, under those salubrious environs, our commander threatened to visit the battalion two days later. Always a pleasure to receive a good soldierly commander, and ours, Brig Rathore, Vir Chakra, from the Grenadiers was one of such class.

A good non vegetarian meal with chilled beer is part of the package for a VIP visit in high altitude. Normally, they are all part of high altitude rations, but the weather had disrupted the delivery of supplies. Getting all this at short notice, in a remote high altitude area was a logistics exercise. A multi directional 'search and seize' mission brought home four chickens, appearing as if they had one leg in their graves. Inquiries revealed that in the entire Budh Karbu village, they were the last four birds, barely standing.

Captain Tarun Kaul, the Regimental Medical Officer (RMO), was pressed into service and ordered by the second-in-command to ensure that they survive till the commander's visit. A spirited and a competent doctor, Tarun took off for his first encounter as a vet without complaint. A resuscitation operation- a few fluids and a healthy diet prescribed by Tarun, to be given under the watchful eyes of the Nursing Attendant brought hopes that they would survive the night to be slaughtered the next day.

Late at night, the commander's visit was postponed due to exigencies of service. Fresh dates for the visit were to be intimated shortly. Another crisis loomed large. Tarun was summoned again and tasked to make the chickens last till the D Day, the day itself not known. Tarun made an emergency visit to the mess poultry, which consisted of two rum khokhas (cartons) with some sacks and old blankets to keep them warm. He reviewed his medical support plan for the long term survival. He returned satisfied that the important mission will be accomplished and gave the situation report to the second-in-command, whose future seemed to hang on the thin drooping necks of the chickens. The Mess Havaldar was instructed to preserve the chickens till the commander's visit, at all cost.

The next day, to everyone's surprise, they found 'chicken curry', the commanding officer's favourite dish, served for lunch. Mess Havaldar Khandpal was summoned and asked, "Why was chicken prepared despite instructions to the contrary?" He replied that he had no option but to do so. During the inspection of chickens along with Mess Cook Bishamber, they had discovered that this chicken was on 'DI List'(Dangerously ill List). Better eat it, rather than lose it. They had taken the joint command decision and managed to slaughter the chicken just before he could breathe his last. Good decision. Surprisingly the dish tasted real good. The cook's explanation was that the chicken got marinated well from inside. We did not understand him then. Later, it was learnt that between the two they had tried to revive the sick chicken, and their master drug was a peg of rum, thrust down the throat of the chicken with a syringe. Unfortunately, that wonder drug did not work.

Tarun saw his job under threat from this team of mess Havaldar and mess cook, who between them decided the life and death of chickens. Finally, all the chickens went through this cycle before the commander's visit could fructify. The cycle started again when he finally visited the unit.

The Indispensable Bukhari and Turlum

- *Sahayak- Officer's batman who looks after his administration, helping him to perform his duties without encumbrance to bother about his personal matters.*

Bukhari is a winter mate that shares an officer's tent. It can mean the difference between life and death in high altitude. Ask any infantry officer, and he will choose bukhari over his wife. It gives him warmth at the turn of a tap, with no more than just a hint of a light. The sahayak prepares it for the officer, gets hot much faster, remains so for as long as you like, even when switched off, barks once in a while, but rarely bites. But when it does, it could be as bad as the lady wife's- fatal.

You can cook almost anything on it for any length of time, it never complains, and works as a night lamp. For the morning jobs or for a bath, you can boil water in a metal container called 'turlum'. It can also be used as smoke generator to guide helicopters. It stays where it is put and never questions where you had been. It is looked after by the sahayaks, costs very little to maintain and survives entirely on military supplies.

Like a wife, bukhari also needs to be looked after. If not maintained, it can bite with a back blast that can transform the innocent looking bukhari lid into a deadly projectile and bring down the tent in a blaze in no time.

Well, I mentioned turlum. A newly raised battalion was inducted into high altitude area in Sikkim in the early eighties, on its first high altitude tenure. The commanding officer (CO) of the unit, himself first time in high altitude, visited our unit and was thoroughly impressed by the utility of the marvel called turlum. Armed with this knowledge, he directed his battalion Quartermaster (QM), Captain Jung Bahadur, to procure one turlum for each bukhari in the unit in next three days. "What is it used for?" asked the QM. "For heating water in a bukhari," replied the CO.

There was no sign of turlums for the next seven days. The CO summoned Jung Bahadur and asked him the reason for the delay. Jung Bahadur replied, "Sir, for the last three days I have scanned the entire lists of Ordnance and EME stores but have not been able to locate turlum, and without the cat part number of the item, I cannot demand it. I will be going to the Ordnance depot tomorrow to get the cat part number of turlum and place a demand," replied Jung Bahadur innocently. The CO smiled. There was no way Jung Bahadur would ever find the cat part number of turlum in his life. Turlum is not an ordnance item. It is a cylindrical metal packing case of a 105 mm artillery gun round, returned to the ordnance depot as scrap after firing. It perfectly fits through the lid space and hangs inside bukhari, boiling water in no time. Having understood, Jung Bahadur managed a truck load of turlums, creating storage problems. The men were seen using turlum for the nature's call instead of rum bottles. Turlums are used for a variety of purposes- decoration in regimental colours, night lamps, route markers in snow, flower pots and a lot more.Innovation thy name is army.

Guardsman Badshah Khan

Guardsman Badshah Khan, walked on four legs, stood taller than the shortest officer and was more handsome than a few. Standing on hind legs, his head crossed seven feet.

A Great Dane was procured for twenty thousand bucks when he was 21 days old and christened Guardsman Badshah Khan. Soon he became the darling of the battalion. He was trained in daily routine. As a small pup he ran behind the PT squad yelping in protest, clearly not liking the idea on a cold morning. PT is a compulsory parade and no one is excused. Guardsman Badshah Khan grew up under strict military supervision.

Soon he was sent for his young soldier's course to a dog trainer in Chandigarh. Six weeks later he returned with distinction. He learnt to receive the VIPs with a hand shake, accompany the Commanding Officer on his rounds, do PT display with the battalion PT display team- jumping through the ring of fire over the high horse. And the grand finale; when the battalion shouted the battle cry, Badshah would stand on hind legs and bark in unison with the cry of *'Garud Ka Hoon, Bole Pyare'*.

Decked in regimental Red and Gold custom made ward robe, he became an icon in Jammu. As his aura spread, officers came to see him from far and wide.

Badshah was trained in guard and patrol duties, albeit on a leash. He was too big and strong to be let loose. Seeing a cat or a dog he would go wild. Bloody aggressive dog, he would tear apart any one if he ever got loose, we used to wonder. God help the terrorist who dared into the unit lines. The battalion was proud of him.

For his sustained outstanding performance, he was promoted to Lance Naik in a proper ceremony. The soldiers would joke how Badshah was senior to them.

While on patrol, he had to be held by two soldiers with a solid steel chain. His look itself was more than a nuclear deterrent. To increase his deterrence effect, a short soldier walked beside him. Badshah's head almost levelled the jawan's head. Civilians around the unit lines would pour out to see him. Dogs and cats seeing him would scramble for cover. Badshah would get restless pulling wildly at his leash, making it hard for his handlers to manage him. *"Wonder, what would become of these dogs and cats if he ever got loose,"* thought the patrol leader.

An incident of a terrorist break-in had happened in a unit in the station. The unit intensified patrolling around the perimeter.

Taking a different route one day, the patrol had not gone too far from the unit, when, a biggish cat sitting under a cot, growled angrily as Badshah Khan passed by. All puffed up, it chose to hold ground. Badshah growled and became restless, threatening to break away. The two handlers struggled to keep him under control. As the luck would have it, the neck strap broke letting him loose. As the patrol stood wondering at the fate of the cat, Badshah Khan bolted off as the street dogs scurried for cover.

The patrol continued on its mission as the dog handlers were dispatched to find Badshah.

Badshah was found hiding under a staircase in the unit, panting heavily, scared to death. To our horror we learnt that the big Badshah Khan was scared of a cat. Fortunately, the incident was revealed to no outsider and remained an in house secret.

It was then a unit mission to guard the secret and retain his deterrence. Someone rightly said, *"It is not the size of the dog, it is the amount of fight in him that matters."*

He grew up too protected. Even a domesticated lion finds his first hunt a dilemma. Do you remember Elsa from the famous

"Born Free" getting thrashed by a small hog. Unfortunately, Badshah died of sickness before we could celebrate his sixth birthday. His demise created a void in the battalion. Everyone missed his pranks on PT and games parade. He was a big source of happiness to the men in the battalion. Lance Naik Badshah Khan was bid farewell with a heavy heart and buried with full honours.

Principles of Management do not Change. Even with Pigs

The units these days get a fair sum of money to meet various requirements. In the earlier days, it was not so. Money for troop welfare or running institutions like the bands, family welfare centre etc was always at a crunch. The units in the army would run a variety of institutions to generate additional income for the welfare purposes. These included running a CSD canteen, a poultry farm, a fishery and sometime a piggery. Now there are orders on what can be done and what cannot be done.

My battalion maintained a piggery because of its class composition, i.e. hill troops and majority from the Northeastern states. Pork was their preferred meat, not to mention that it was a profitable venture and provided the much needed additional funds to run the unit.

So, there we were, a well maintained and a hygienic piggery with massive pinkish Yorkshires. As the second-in-command I was responsible for all such institutions. I knew that a female pig, named Hema, was pregnant and there was excitement in Havaldar Kishan Singh, the senior safaiwala who was responsible for the piggery. I would get a daily report on the developments.

One day, I was surprised to see Kishan come to my house in the evening. That was quite unusual. He looked very happy and excited. So I asked him what was up.

He said, *"Saab, sub theek hai. Sri maan, main aapko khush khabari dene aya hoon kee suari 'Hema' ne dus bache diye hain. Ismen se paanch ladies and aur paanch gents hain. Ap ko bahut bahut mubarak ho saab, aap kee hee maharbani hai. Saab Hema aur bache theek hain. Maine unke dekh baal ka pura bandosbast kar diya hai. Aap kal wahan ayie mein apko detail mein brief karoonga."*

(By God's grace, all is well. I have come to give you good news. 'Hema' has delivered ten piglets. Heartiest congratulations, it is all your doing and all the credit goes to you. Out of ten piglets, five are ladies and five are gents. I have made all arrangements. I will brief you in detail when you come there tomorrow).

I was impressed by his enthusiasm, and was amused when he congratulated me on the birth of the piglets. I swear I had no hand in the process.

I reciprocated his compliments and congratulated him as well. In fact, he deserved all the credit for he had looked after half a dozen pigs with devotion.

I went to the piggery the next day to see the new arrivals. I could not see any, and then Havaldar Kishan took over the proceedings. He removed a hessian cloth and there were two rum khokas. Each contained five pink cute looking piglets, no bigger than mid sized rat, wriggling with their eyes closed.

"Saab jaise maine kal bataya tha, total dus bache hain, jisme, paanch ladies aur paanch gents hain. Command aur control ko madye nazar rakhte hue maine in dus bachon ko do section mein banta hai. Ek se paanch tak Number 1 Section aur chhe se dus tak Number 2 Section. Kyonki suari let kar dood pillati hai, at a time sirf paanch bachhe dood pee sakte hain. Is liye maine donon groupon ko alag- alag time mein banta hai. Har group ko

ek baar mein 15 minute ka time diya hai. Is liye kee chatak bachhe dobara number na le len, pahchaan aur command control ke lie maine Number 1 Section ke mathe par kala teeka lagaya hai. Pahchaan ke ilawa inko kisi ki nazar bhee nahin lagegi. Alawa is ke, inme se sabse kamzor bachon ko Number 1 Section men rakha gaya hai, jinka dood peene ka number sabse pahle diya jata hai."

(As I told you, there are ten piglets, of which five are ladies and five are gents. For the purpose of command and control I have divided them in to two sections. From one to five is Number 1 Section and from six to ten is Number 2 Section. Because the mother pig lies down on her side to feed the babies, only five tits of one side are exposed. I have therefore divided the time between two sections. Number 1 Section feeds first and each section gets fifteen minutes at a time. To ensure that the smarter ones do not bully the weaker ones, I have put all the weaklings in Number 1 Section. For the purpose of identification I have numbered them from one to ten, and marked all the members of Number 1 Section on the forehead).

He then carefully picked up Number 1 Section and lined each one with a tit of the mother pig. I noticed that apart from the 'kajal' on the small forehead, he had numbered them from one to ten on their small butts. I clearly understood why, they all looked the same.

I was thoroughly impressed by his foresight, management and execution of an otherwise mundane job. I complimented him for his work and he got his best compliment from God, all the ten piggies survived. He looked after them as if they were his children. He said, "Saab, I have only six months to retire, please do not kill them for pork till I leave the unit." I promised him that and kept that promise.

High Delegation's Visit to the Battalion

We were deployed in North Sikkim in 1980 for the operational familiarization. The whole battalion was in tents in that remote high altitude area. The corps commander accompanied by the divisional and the brigade commander decided to visit the battalion at a day's notice and stay overnight.

Three top commanders would be a handful anywhere, but when they have to be accommodated in tents at short notice in high altitude, they were more than a bucketful. Our commanding officer, the inimitable Col Pup Mann (Later Lieutenant General) took a small conference as we sat for dinner in the tent. He said, "We will look after them as well as we can, with what we have under the circumstances. I know you will look after the corps commander and the divisional commander, for they are big men, remember, the dog knows his master and not his master's master. Do not forget your brigade commander."

The next day the cavalcade arrived. They were put in separate 180 Pounder tents. Those were the days of magneto telephones when all the calls had to be taken through the exchange. The Mess NCO, Havaldar Khandpal asked the exchange to be connected to the second-in-command. He then said,

"Saab, mein Mess Havaldar bole raha hoon. Saab achanak itne sare senior saab log agaye, yahan high alti mein, na toh koi chadar, razai aur takiye hain aur kisi karan se ration bhi supply se nahin aya, yeh to bari bejatti ho jaegi saab, ab kya karen." (Sir, I am Mess Sergeant speaking. So many senior officers have landed up suddenly in high altitude , creating whole lot of problems. We have no beddings, quilts etc. To add to the problems, even the rations have not been delivered today. We will lose face this way. What should we do?)

"Koi phikar karne ki batt nahin hai. Achanak saab log ayenge

toh aisa hee hoga. Kuch nahin hai toh anda bhujia hee khila dena." Mess Havaldar was surprised to hear the reply. (There is nothing to worry, if they come suddenly at short notice then this has to happen. If you have nothing then serve bread and eggs).

"Arre saab, corps aur divisional commander ko anda bhujia se to nahin tarka sakte, kuch to karna hee parega, mein dekhta hoon." (Can't dispose the corps and the divisional commander with bread and eggs in dinner. Will have to do something, I will see what can be done).

"Koi dekhne kee jarurat nahin hai, mein General Balaram, apka corps commander bol raha hoon," said the corps commander. (There is nothing to see, I am General Balaram, the corps commander speaking).

The corps commander happily narrated the incident later and appreciated the NCO's concern.

5 GUARDS Unfit for Peace?

5 GUARDS, as I said before, is a pahari paltan with its troops belonging to hill regions from the entire spectrum of Himalayas from Kashmir to Manipur, a fantastic troop composition. The battalion has fought all wars since raising with distinction and done us proud. It has one shortcoming though, it fails miserably when you ask it to do anything wrong, steal for example. A review of some memorable experiences.

The armies are always short of resources. Manage is the mantra. The units manage to accomplish the task, no matter what. Our battalion was doing collective exercise in deserts.

In deserts, with stretched frontages, all units were short of signal cable. There was a major problem of loss of signal cable laid to provide communication to units. Huge chunks of signal cable was disappearing every second day. The commander, realizing that units were the culprits, conveyed his displeasure to the commanding officers in a conference. Colonel Pup Mann, our commanding officer (CO) assured the commander that 5 GUARDS could never do such a thing. The commander nodded in agreement.

Next day the commander visited our unit. Col Pup Mann was driving him around the exercise area. As they drove over the dusty camel track, Pup Mann noticed some fishy movement behind a bush a hundred yards ahead and recognized a familiar silhouette. He hoped the commander had not noticed. As he approached that point, the commander said, "Just stop here Pup, I think I can see someone hiding behind the bush." Pup stopped the jeep fearing the worst.

The commander looked out and lo and behold, there was Lieutenant Harminder, the Signal Officer of the unit, on a special mission, crouching behind a bush with a wire cutter in his hand 'managing' signal cable for the unit. Harminder, a Sikh officer with mischievous eyes has such innocent looks that no one would believe what he was up to. Looks can be deceptive, as they say. The commander looked at Pup Mann, saying nothing. As for Pup Mann, he had assured the commander no such thing ever happened in his unit, but took it in his stride in good humour, as he often absorbed the 'goof ups'. Well, in the units junior officers sort out minor problems at their level and do not bother their commanding officers. Colonel Pup Mann, later a corps commander, often used to narrate this incident zestfully.

Colonel Pup Mann was a fine CO and I was privileged to be commissioned during his command. He always encouraged young officers and exposed them to senior officers to build them up.

During the same exercise, he was travelling in his jeep with Second Lieutenant KC Sadyal, the battalion Intelligence Officer. Unlike these days when majority of the officers own cars, those days most young officers did not know driving. Pup Mann asked him if could drive. Sadyal said, yes he could. "Come on, drive, " said the CO.

Sadyal awkwardly slipped in behind the wheel, started the vehicle and rolled on. Hardly a driver worth the name, the vehicle coughed, stuttered and leaped and jumped as Sadyal juggled with a combination of gear, clutch and accelerator. Pup Mann watched him from the corner of his eye, wondering what he had landed in to. The vehicle slowly settled into a smooth ride. True to his style, he had put his life in danger and sat quietly as Sadyal drove on like a robot. A while later, Pup saw the brigade commander standing next to his vehicle on the road side. The commander also signalled Pup to stop. Pup asked Sadyal to stop next to the commander's vehicle. As the vehicle approached the commander, Pup noticed something was amiss, the vehicle was not slowing down. "Stop here," said Pup. The commander saw the vehicle heading straight for him. He too realized that something was not right. He dived for his life, just managing to get out of the way as Sadyal rammed into the commander's jeep.

Commander was not amused as he got up and dusted himself, neither was Pup, but he himself was the architect of this episode. Sadyal had almost written the commander's obituary, brought an end to Pup's bright career and dug his own grave. The commander said, " Pup, I have travelled 1200 kilometres from Meerut only to be killed by my own battalion." Pup looked at Sadyal enquiringly with a smile, Sadyal honestly said, "Sir, I only know how to start and move a vehicle, I do not know how to stop it." Pup in his inimitable way said, "Sir, no bad intentions, only a navigational error. I told him to drive; only now I realize Sadyal is a danger to own troops on the road." The incident was talked of for long in good fun.

Later our battalion moved to Siliguri in 1979. We were in East Sikkim for area familiarization. One Sunday, all officers came to Gangtok for a meal out. The main street had good looking lamp shades on the lamp posts all along the street. The CO, Pup Mann casually remarked, "Impressive lamps, will look good in our officers' mess lawn."

Three days later, Pup Mann was driving back after a conference from the divisional headquarter at about eleven in the morning. To his horror, he noticed two well decked Guardsmen in uniform hanging from the lamp post, removing the lamps in the middle of a crowded street. He stopped and called them over. He was told that they had been asked by the adjutant to remove five of these lamps for our officers' mess. Pup Mann realized the genesis of the problem. And who was the adjutant, our faithful Captain Harminder. The hanging Guardsmen were loaded into the CO's vehicle and brought back to the unit before they could do any harm.

Lesson learnt: officers in position of authority need to be careful about loud thinking or casual remarks. Some subordinates will act even when action was not intended.

The Smart Allec JCO

In 1994 my unit had moved to Bhatinda after a long tenure in the valley. We were sharing accommodation with another unit. The accommodation was very tight and temporary structures were created for unit institutions. Coming back from the Golf course, our officer saw a huge uprooted eucalyptus tree. That is too good an asset to let go, he thought. That would be useful for many things in the battalion and a night operation was planned to retrieve the tree that very night.

We picked our smartest go getter JCO for the job, Subedar Ganesh Dutt, an outstanding sportsman. He was briefed in detail on execution of the plan and sent on a reconnaissance. He reported that the tree will have to be cut into three logs to fit into the vehicle. The troops to task was accordingly worked out and the party was suitably equipped.

To achieve surprise, the NMB (No Move Before) was selected as mid night. It was felt that by then all the movement would have finished.

The party got to the spot and got busy cutting the huge tree. As the luck would have it, the commanding officer of Provost Unit (Military Police) was returning from a late night party in his official gypsy. He saw a 3 Ton vehicle on the roadside and soldiers frantically cutting the tree. The officer stopped and called for the party commander. Subedar Ganesh reluctantly went to him.

"Kya kar rahe ho?" (What are you up to?) asked the officer.

"Hum yeh gira hua ped le ja rahe hain," (We are taking this fallen tree). replied Subedar Ganesh.

"Kis ki permission se?" (With whose permission are you taking this tree?)

"Permission toh kisi kee nahin hai saab, yeh toh toota ped hai." (We have not taken any one's permission. This is a fallen tree).

"Toote ped ke liye bhee permission chahie hoti hai. Aap kis battalion ke ho?" (You need permission even for a fallen tree. Which battalion are you from?)

Subedar Ganesh did not want to divulge his unit so chose to remain quiet.

"Maine poocha saab aap kis unit ke ho?" (I asked which unit are you from?) repeated the provost officer firmly.

Unable to wriggle out, and no way he could fool around with a provost officer, Subedar Ganesh thought quick and formulated what he thought was a smart reply.

"Hum 5 GUARDS ke naheen hai saab." (We are not from 5 GUARDS sir).

The Provost Officer drove away without a further word. The next day he called our commanding officer, who was known to him and narrated the incident. He advised that if you send someone on such missions, at least send some smart allec. The commanding officer could not tell him that he was our smartest, especially selected for the special mission.

Believe it or not – Surprise all Around

Lieutenant Dal Bahadur, a short roly- poly Gurkha was a wonderful bloke. Having been commissioned from the ranks, he already had two children at the time of commission.

He joined the unit alone after commission and stayed that way for a long time. He then proceeded on two months leave. It was a Saturday afternoon and we all had collected in the mess for a weekend beer. There sat Dalle, as he was affectionately called, back from leave with a smile of happiness.

"Good to see you, Kancha, welcome back home," I said shaking his hand.

"Thank you, sir," replied Dal Bahadur with his wonderful smile.

"Sir, Dalle has got some good news," said Yogi Paul.

"And what's that Dalle?" I asked.

"Sir, he has been blessed with a baby," replied Yogi Paul, before Dal Bahadur could answer.

"That's a bloody surprise Dalle, you never told us that was coming," I said.

"Sir, it was a surprise for me also. I also did not know it was coming," replied Dal Bahadur with his beaming smile.

The Fish Shall not Die of Old Age

Troops love to be out on training for it keeps them away from meaningless peace time rigmarole and they do what they are meant to do. We were out on company collective training near Birdhwal in Rajasthan. Bulk of the troops in my battalion are from the Northeastern states. Happy go lucky by nature, they simply love mutton, fish, chicken and pork. Accompanied by a tot of rum, it makes their day. Customarily, it is a given in the battalion that 'meat issue day' is accompanied with rum issue.

As we sat in the training area with everyone busy on his squad post, Havaldar Basumatari, the ever cheerful Company Quarter Master, walked up to me, *"Saab Suna hai, idhar men, gaon ke paas bahut talab hain. Usme suna hai bahut macchi hai saab. Yeh Rajasthani toh khata nahin hai saab. Bechara macchi burhapa se mar raha hai. Saab, Yeh to theek nahin hai saab,"* said Basumatari with all the seriousness. (Sir, I have heard there are ponds near the villages here. These ponds contain huge amount of fish. The locals are vegetarians and do not touch fish at all. The fish are dying of old age, what a shame).

"Toh kya rai hai apka?" (What is your suggestion then?)

"Saab, macchi ka burhape se marna acchi baat nahin hai. Apka izzazat ho toh ham ja kar macchi pakar ke laiga aur pura company ko khilayga," said Basumatari. (It's not good for the fish to die of old age. If you permit, we will go fishing and the whole company will have fish).

"Theek hai jao," (OK, you can go), I said.

The evening was a feast. The fishing team had returned with a vehicle load of massive fish. The JCO came and suggested I should have a drink. With fish fried galore, it was a wonderful evening under a Keekar tree (scraggy desert tree) next to the quietly flowing waters of Bikaner canal.

Come the next day, and there stood Basumatari again with a smile. Without his saying a word, I said, you can go but after the training is over. A bit dejected, he walked away. The evening was a repeat performance, the company having a field day with rum and fish.

The next day Basumatari was there again and when I told him he could go after the training was over. He stood there silently.

"Anything?" I asked.

"Haan Saab. Saab yeh training toh hota hee rahta hai. Lekin company aisa machhi kabhi nahin khaya saab. Yeh moka kahan milta hai saab. Agar apka izzazat ho toh main abhi jaoon aur phir lunch aur dinner men machhi hoga saab." (Yes sir, this training is fine. We do that all the time. Won't get a chance to eat fish like this again. So, if you permit, we can go fishing now and then the company can have fish both for lunch and dinner).

I realized what fish meant to an Assamese and saw the sense of his logic, from his point of view. I let them go every day.

The company had a total ball for weeks and then the senior JCO walked up to me with Basumatari in tow.

"Saab, yeh machi parade band karna parega saab." (We will have to stop this fish business).

"Kyon?" (Why?) I asked.

"Saab, yeh machhi ke chakkar men sara tel khatam ho gaya hai saab. Ab baki khana banane ke lie tel nahin bacha hai." (Sir, we will have to stop this fish parade. Because of this fish cooking business, the company has run out of cooking oil. No oil left to cook company's meal).

Genuine problem I thought and then looked at Basumatari.

Basumatari replied *"Larai ke halat men bhi kabhi tel khatam ho sakta hai saab. Company ka rai hai ki machhi band nahin hona chahiye. Tel ka problem hai toh aj se har jawan ko personal machhi issue karega saab. Woh individual khud apna tel khareede ga aur banaega."* (The oil can finish in battle also. The solution is that henceforth I will issue fish to individual soldiers. They will buy their personal cooking oil and cook themselves).

That was a fair one, I thought. Realizing how much fish meant to my men, I readily agreed and then the men managed themselves, but the fish feast continued.

After about four weeks, they had scanned every pond within a hundred kilometres with a vehicle camouflage net and finally the fish had dried. The company had returned to normal diet.

Basumatari was an excellent soldier, extremely talented and full of life. He could play a few instruments, sing well and play any song with a leaf plucked from the nearest bush.

One evening, taking a walk along the Bikaner Canal, I saw someone standing motionless in waist deep water in the canal. He had a bamboo pole held over his head, ready to strike. As I stood and watched, the man swung the pole splashing lots of water, and in one smart move dipped his hand and came up with a good size fish in his hand.

I was impressed. When I went closer, it was good old Basumatari on his fish hunt. He had already collected a few. His fishing technique, archaic though, was effective, and his love and determination for fishing undiminished.

To be or not to be

Amar Singh, an excellent JCO from 1 GUARDS was commissioned as a Regimental Commissioned Officer (RCO) and posted to my battalion. He had been in the battalion for about a year. One day, gossiping outside our rooms in Khrew, he said, “Sir, in the last one year I have been drinking too much.” “How come?” I asked .

He replied “The first six months in the happiness of becoming an officer, and the last six months, in the sorrow that I had become one.” He said, as a JCO, there was not one year when he did not avail his full leave. As an officer, he had not gone on a day’s leave, and was unlikely to get any in the remaining year. Not easy being an officer.

Much later, Amar was deployed with his company in an active area where cross- border firing was the order of the day. There was a rumour that an officer of the neighbouring unit was involved with some local woman. I rang up Amar and enquired if he had heard about this. He said, yes he had heard about it. I then asked him, “Hope you are not involved in such extracurricular activities in a sensitive operational area.” He replied in his broken English, “Sir, at my age, weight and service, it is all null and void.”

He was six foot three, weighed one hundred and twenty kilos and was fifty four years young. I guess he did not want to take chance of being charged with ‘culpable homicide not amounting to murder’.

Later, during an exercise involving a long endurance march in the mountains, I was worried that Capt Amar may find it difficult. To my pleasant surprise, he had managed well. I told him he was much fitter than he looked and had done well.

“No sir, I am hardly fit. I had told my batman to hang a rum bottle from his rucksack and stay ten yards ahead of me. That was the secret of how I am here and the bottle now is empty," replied Amar in his broken English. "And yes, I have another bottle for the way back."

Foren Language

A Sikh officer joined us in Srinagar. Coming from a village, he needed to be worked upon on various aspects of his personality to fit into the unit groove. I was the adjutant and noticed that he was always listening to bhangra and noisy Punjabi music. His English was certainly via Bhatinda.

In the long term interest of the battalion and the officer himself, I started working on him. I confiscated all his Punjabi cassettes and ordered him to listen only to English music. He was instructed to walk, talk, sleep and dream in English.

He asked me if he could bring his motorcycle from Gurdaspur to the unit. We allowed him to do that. A few weeks later his black Bullet motor cycle was in the unit and he was proudly cleaning it.

I said, "Great, Pammy, how did this bike get here?"

He replied, "Sir, from Gurdaspur to Jammu by truck and then by foot." (He had driven from Jammu to Srinagar).

No Sir....Says Jimmy

A young officer joined my battalion from the Officers Training Academy, Madras in 1985. We decided to call him Jimmy. Going through the list of officers commissioned with him, we learnt that he had passed out 'second-last' in order of merit in his course. That was not encouraging since we had two 'Sword of Honour' awardees in the paltan. "Never mind, we'll make him one in the battalion," we thought.

We were located at Khreuh near Srinagar and I was doing the adjutant of the battalion. The brigade was given a courier duty from Srinagar to Udhampur. One officer each was to be detailed from my unit and the neighbouring unit, 8 JAK LI. Jimmy, a Texan by descent was a wonderful ever smiling character with chubby cheeks, prominent ears and had a loud infectious laugh. He was detailed on the courier duty. Courier duty involved taking a sealed box with classified documents with an armed escort under the responsibility of two officers. Being his first courier duty, I briefed him in detail on what all it involved and what he needed to do to deliver the documents safely.

Having briefed him, I said, "Even though I have briefed you in detail, you don't really have to worry. There are two officers detailed from the brigade on the courier duty. Given your lousy passing out order of merit, the other officer has to be senior to you. He will be the officer-in -charge and primarily responsible and you can generally relax."

"No sir," quipped Jimmy enthusiastically, "The other officer detailed for the duty is my course mate Second Lieutenant Mahwa from 8 JAKLI. He was last in order of merit."

Both of us had a good laugh. Everyone laughed their guts out in the mess that evening when I narrated this exchange.

Whose Backside is it Anyway

I was doing the adjutant of the battalion in Khreuh near Srinagar. Young Jimmy coming from a Texan environment wasn't particularly friendly with English and was not conversant with the army lingo yet. He had hardly served three months and it was time for the annual inspection by the brigade commander. This is an important event for a battalion, as the formation commander reports on the battle preparedness of the unit based on this inspection. The troops accommodation was in double storey barracks, in which, one company occupied the ground floor, and the other-the first floor. A lot of effort was put in to clean the area along with military spit and polish, the so called 'Geru-Chuuna'.

A day prior to the inspection, the commanding officer, Colonel Umang Seth went around the battalion. He walked in to my office and said that he was not happy with the cleanliness of Charlie Company area and wanted that sorted out. As it often happens, the company commanders at critical times are invariably the subalterns. The company commander in question was the green horn - my friend Jimmy. His company was occupying the ground floor. I called him to my office and the following conversation took place.

Self : "The commanding officer went around the unit lines?"

Jimmy : "Yes sir."

Self : "I believe your company was ticked off by him today for poor cleanliness."

Jimmy : "Yes sir."

Self : "What's the problem? Why can't you bloody well keep the damn company area clean? You know the annual inspection is around the corner?"

Jimmy : "Sir, there is a problem, sir."

Self : "What is the problem?"

Jimmy : "Sir, the problem is that there is a lot of confusion about the responsibility for my backside."

Self : "What is the confusion?"

Jimmy : "My company is on the ground floor and Delta Company is on top of me. My responsibility is to clean my front side and Delta Company's responsibility is to clean my backside."

Self : "Why can't you clean your own backside?"

Jimmy : "Sir, because it is too large with a lot of wild growth all over." He replied sincerely.

Self : "So what is the solution?"

Jimmy : "Sir, the solution is that my backside should be divided into two halves. The cleanliness of the left half should be the responsibility of Charlie Company and the right half that of Delta Company."

That left me with the obvious question, "Who will clean the middle?" He did not understand, "I will do it myself," he replied sincerely.

Well, Jimmy is one of the finest officers I have had the privilege to serve with. He passed out of OTA below par, way down in merit, not playing games etc. Highly spirited, mentally rock solid, he developed into a highly dependable officer. Given a task, it was taken as done. If he came back and told you it could not be done. You knew that he has tried everything less murder and that it was not worth pursuing further.

A solid positive frame of mind with solid regimental spirit, he was the man to pick for the worst job. He would give it everything, even his life, and remember he had passed out at the bottom of his course. The academy merit, he proves is of little relevance. The tougher the going got, the more he smiled. It is unfortunate that these wonderful soldiers, who will win wars for the country, are left out in the march to higher ranks.

Russo- German! Does it Matter?

Sports have a special significance in the army. Regular inter unit competitions are held in most of the sports. Each unit has certain favourite games depending on the troops it has. The units ensure that those sports competitions are won, while they attempt to do well in others. The brigade organized an inter battalion wrestling competition. My battalion with its hill troops had no inclination for wrestling. The battalion had in fact never participated in a wrestling competition before and thus did not have any wrestlers. The other battalions were Dogra and Rajputana Rifles where wrestling is very popular. With great difficulty we managed to get out of the brigade wrestling competition. The battalion was however asked to provide one officer as a wrestling referee. All our efforts to explain that no officer even knew the ABC of wrestling fell on deaf years. Nothing doing said the Brigade Major, you will provide a referee.

In the army, there is not to reason why. In the unit it was a problem to detail a referee for wrestling as no one had the foggiest idea. As always, it is the junior officer on whom the axe falls in crisis. Our 'man Friday', Jimmy, whose latent surname was Dahiya, was nominated as referee because of his following credentials :-

- He was a Texan with roots in Haryana where wrestling is a popular sport.
- He was a youngster and could not protest.
- He looked closest to a healthy wrestler with chubby cheeks and prominent ears.

Jimmy, as he was known, tried to plead his case stating he had no idea of wrestling and will only make a fool of himself. His representation was disregarded with disdain. He was asked, "You are a Texan, your father must have wrestled some time in life." "No sir, he is a government servant," Jimmy replied.

"In that case your grandfather must have definitely wrestled." Taken aback, he said, "Yes sir, he must have wrestled." His genetic connection with wrestling was established and having been found closest to wrestling was nominated as our referee.

Jimmy attended the coordinating conference at the brigade headquarters. I met him as he was walking back to the unit.

"What happened in the conference, Jimmy?" I asked him.

"Sir they realized I am useless for them, so I am not required now. They have decided to have 'Russo-German' type of wrestling," he replied.

"What kind of wrestling did you say?" I asked.

"Sir, Russo-German wrestling."

"Are you sure?"

He thought for a moment and said, "Yes sir, very sure, Russo-German style of wrestling."

Then I said, "Russo-German is the Second World War campaign between Russia and Germany. May be the wrestling is *Graeco – Roman.*"

Jimmy thought for a while and burst out laughing in his typical style and said, "Yes, yes, it is not Russo-German, it is '*Greeko-Roman*' sir." We laugh over these episodes whenever we meet.

Fortunately, the brigade headquarters left us alone.

Uneasy lies the Head that Wears the Crown

The battalion was located at Udhampur. During a monthly Sainik Sammelan (Battalion Commander's address to the troops), I was surprised to see my commanding officer (CO), Colonel Seth, sweating profusely while addressing the battalion, while everybody else sat pretty. Surely, he couldn't be nervous in front of the troops, I thought. Looking up, I saw a fan right over his head in full glory that further increased my anxiety. Why is the CO sweating when the whole battalion, also under fans, was perfectly comfortable, I wondered.

A little later the power failed, the fans slowed down and gradually stopped. When I looked at the fans, the one over the commanding officer's head was rotating in the opposite direction blowing the breeze upwards. Of all the fans in the hall, it had to be the one over the commanding officer's head. Uneasy lies the head that wears the crown. Trust the MES to do the unthinkable.

Soldiering at its Best

Soldiering at its best was witnessed during a party hosted by Colonel Ranjit, Commanding Officer (CO) 5 GUARDS. A Sikh company commander, Major Dalbir Singh, was an excellent officer who liked his daily drinks. Sometimes, after drinking, he would get sentimental on soldiering issues and get into a lecturing mode to a captive audience (junior officers). One day earlier he apparently had a quarrel with his wife. The matter came to the notice of the commanding officer's wife.

A day after the incident, the battalion was celebrating

its victory in the formation Basketball championship in the commanding officer's house. I was chatting with the CO when his wife informed him that Major Dalbir had fought with his wife, and that he should be suitably counselled. From the corner of my eyes I saw Dalbir standing close by and thought he would have heard the lady complaint against him. I was right, he walked up to us and said, "Sir, Mrs. Ranjit is complaining to you about my fight with my wife. So what is wrong, I am a soldier and fighting is my profession." Colonel Ranjit had no answer. He laughed his heart out as did the rest of us and simply said, "Dalbir, you are a great soldier, and I admire you for it, but I do not want you to take your profession so seriously." The matter finished in good humour.

Dalbir was indeed a devoted soldier. A task given, it was all that mattered. He never crossed the line after drinks; mostly became quiet and a silent observer. There was one clear indicator that he was happy. It was the pronunciation of the word 'tick-tee-boo', a word he used often. As long as he was pronouncing it correctly, he was fine. The moment he started pronouncing it as 'tick-bee-too' rather than 'tick-tee-boo', we knew he was happy.

We had a highly decorated commander from Grenadiers in Khreuh, Srinagar, in 1987. A simple down to earth soldier, everyone loved him. Coming straight from the bottom of the heart, Dalbir expressed his affection one day by simply kissing him during a social function, saying, "We love you, sir". The commander smiled and said, "Thank you Dalbir, I love you too."

I remember, that day he had reached the 'tick-bee-too' stage.

Subedar Major Diwan Singh; A Rare Breed

The Indian Army has a fine institution of the battalion Subedar Major (SM) in the units. The SM is the oldest and generally the wisest of the JCOs. He is respected as much by the commanding officer (CO) as by the other officers and is looked up to by the troops as a father figure. The CO also sees him as a friend, philosopher and a guide on unit matters.

In my battalion, the SM was a veteran by the name Diwan Singh, a simple highlander from Kumaon hills. He was completely illiterate and a typical highlander. He had made his way to this prestigious rank purely by his honesty, loyalty and hard work. Every rank since the rank of Lance Naik (Lance Corporal) was considered his last. But he kept going upwards, finally becoming the battalion SM with the experience of twenty eight years of service behind him. Because of his childlike innocence, he was very popular with the young officers. Upright in his behaviour, unquestionably honest and dedicated, he was extremely respectful to all irrespective of their rank.

Serving on Siachen, we lost three boys in one day due to artillery shelling. The same day we got the news that his teenage daughter had died because of stove burst in his village in Kumaon. As the adjutant, I broke the tragic news to him and told him he can proceed on leave straightaway and I'll inform the CO.

Diwan Singh declared, "With three boys here for cremation, as the SM, how can I even think of leaving. In any case the telegram has reached me too late. By the time I reach home, all ceremonies would be over and I will not be able to see the face of my daughter." He left the day after cremation.

Do you need a highly educated soldier? I would bet my last ten bucks on this illiterate soldier over any IIT graduate.

Diwan's Practical Logic

Located at Chorbatla, Ladakh, in high altitude, Sunday was no different from any other day, except that you could sleep a little longer, feel good with little change in routine and of course, a customary mandir parade.

The mandir parade finished with the religious teacher reciting, what I later learnt was called Durga ji mool mantra - *"Aum-aime-hreen- klim –chamundaye-vichay-durgaye namah."* I have never been very keen on 'matters- mandir', but this one interested me since it sounded very different when recited, a bit funny; resembling Chinese I thought.

Knowing that Diwan Singh would have some good explanation for it, I asked him, *"Saab, yeh mantra bara ajeeb sa lagata hai sun ne mein. Baaki Sanskrit ke mantron jaisa nahin hai. Is kee kya kahani hai?"* (Saab, this mantra sounds very different, not like other mantras in Hindi or Sanskrit, why is that so?)

"Haan saab, aap theek bole rahe ho. Yeh mantra ajeeb sa hai. Asal mein bahut saalon pahle yeh mantra kahin kho gaya tha. Bahut saalon ke baad yeh mila toh China mein mila, aur jab mila toh is halat mein mila. Isi liye sun ne mein Chinese jaise lagata hai." Wonderful explanation.

(Yes, saab you're right. Some years after this mantra was made, it got lost. Nobody knew where it had disappeared. Many years later, it was found in China and when it was found, it was in this shape. That's why it sounds like this, a bit Chinese).

It was an explanation given with supreme confidence and without a second thought. His sincerity and belief in his answer would have fooled anyone, except that I think he genuinely believed what he said.

Diwan and his Supreme Confidence

Colonel Ranjit Singh of 5 GUARDS, later Brigadier, was a big man with even a bigger heart. A good human being and a great commanding officer (CO), he had a terrific memory for troops' names. I guess he knew all the men of the battalion by name, over eight hundred of them. During the opening address of Lance Naik promotion cadre, he rattled out the names of all the jawans. He was bang on-getting all the names right. As the adjutant I had accompanied him, and he asked me to name them, which I did. He then asked Subedar Major Diwan to name the jawans. The Subedar Major (SM) remained quiet for a while as Colonel Ranjit waited. He then simply said that he will not tell the names now, he will tell the names later. When the CO asked him,"Why not now?" He replied in typical Kumaoni accent, *"Shaab, kyonki abhee main inko shtudy kar raha hoon."* (Because presently I am *'shtudying'* them). He had nearly thirty years service then, of which almost twenty seven were spent in the battalion and he was still studying them. We all had a good laugh.

I guess the incident must have been haunting friend Diwan Singh all the time. He must have felt he was not up to the task that day. The CO was a fitness freak and took long walks every evening. Old stalwart Diwan gave him company occasionally. One day the CO was on his usual evening walk with the SM. Enroute they crossed three jawans returning from the water point.

The SM suddenly stopped them. *"Yeh Chokron thum,"* (Soldiers halt), and said, *"Saab, mein aapko in chokron ke naam bataounga. Saab yeh hai Havildar Phullu Gohain, yeh hai Naik Rajkhowa aur yeh hai Naik Dal Bahadur Gurung."* (I will tell you their names and proceeded to name the three soldiers one by one).

The jawans smiled, said Ram Ram saab and walked off. Colonel Ranjit smiled and simply said, "Good saab, well done." Later, during the dinner in the tented mess he narrated the incident with his zestful laugh. The names told voluntarily by the SM, at his own initiative, were all wrong-without exception.

Colonel Ranjit with his big heart never let him know that he was wrong. Subedar Major Diwan's confidence had visibly returned.

Diwan and Field Latrines

The battalion was posted in Ladakh where the troops lived in tents. The Commanding Officer, Colonel Ranjit was very particular about soldiers' hygiene and sanitation. Soon after the battalion had arrived, he decided to inspect the field latrines. As the adjutant, I also decided to accompany him on his round.

In the army, the location of field latrines is indicated by positioning a yellow flag atop a high pole. On reaching the location I noticed that there was a yellow flag all right, but it also had a big black circle in the middle. Frankly, I had not known that there was supposed to be a black circle on the yellow flag. I asked our good old Subedar Major Diwan Singh.

"Saab, aaj tak toh mein samajhta tha ki field latrine par peela jhanda lagate hain, lekin is mein toh black circle hai. Is black circle ka kya matlab hai?" (I did not know that the yellow flag on the field latrines is supposed to have a black circle in the middle. What is the circle meant to mean?)

He replied seriously, *"Saab black circle ka matlab hai bull, matlab bull mein maro. Sara area ganda kar dete hain yeh chokre."* (The circle means bull. It is to remind the soldiers to shoot in the bull, otherwise they dirty up the whole place). In the shooting parlance, term bull denotes the centre of the target. Diwan's practical logic was spot on.

Diwan and the RTC Keys

The Subedar Majors (SMs) in the units are responsible for the safe custody of what is known as the Regimental Treasury Chest (RTC). There is much less in the name than meets the eye. The RTC is a rather heavy unwieldy steel locker weighing over a hundred kilos. It is secured to the wall with strong steel plates. It would need five soldiers to lift it. It has a special number lock and double locks. Additional security is provided by a massive steel chain that would scare an elephant.

The RTC is placed in the unit Quarter Guard, guarded round the clock by an armed guard of eight soldiers.

A look at the multi layered security would tempt any burglar to have a go at it. Surely, with this security it had to have a fortune in it, but would not risk a thing with the massive security apparatus in place.

Actually, twenty nine days in a month, the RTC holds no more than a few hundred rupees, and often less than a hundred. It is only for a few hours on the pay day, once in a month, that the soldiers' pay is kept there till it is distributed to the men. Now with the troops' pay also going to their bank accounts, even that is not happening at most places.

So, frankly, even a petty thief would not contemplate wasting his precious time and energy on RTC, if he knew what it contained.

The SM is the custodian of the RTC who accounts for whatever little there may be in it. The SM holds the original key of the RTC. The duplicate key is supposed to be kept in the duplicate key box kept in the quarter guard, as are the keys of other institutions of the unit. The duplicate key box has a glass front with wire mesh so that the keys inside are visible and can be taken out quickly in case of emergencies like fires. During my routine round to the Quarter Guard as the adjutant, I observed that the slot for duplicate key of RTC was vacant. I asked the guard commander who had no answer. His reply was the standard army reply that he took it over like this from the previous guard commander.

When I asked our old faithful Subedar Major Diwan Singh, he said, "Shaab don't worry, it is shafe." My all efforts to extract the location of the duplicate keys from him met with no success. He assured me that it is in a perfectly safe place which only he knows. He would not part with the secret information, afraid that it would compromise the security of RTC. I tried to convince him that he need not worry about the safety of the keys as there cannot be a safer place than the duplicate key box in the Quarter Guard, but without effect. He just would not budge.

Finally, after much persuasion he agreed to part with the secret of the keys. He took me aside, far from anyone's hearing range and whispered in my ears ***that he had kept the duplicate keys safely locked inside the RTC itself.*** I did not know whether to laugh or cry. I tried to explain the pros and cons of leaving the duplicate keys inside the RTC. Nonetheless, I failed to convince him of the consequences in case he lost or misplaced the original keys by mistake. He simply said that there was no question of him losing the keys. After he showed me why, I believed that as well. He took out a big bunch of keys hanging round his neck from a solid steel chain. The chain never came out of his neck, not even when he was bathing. The only time it came out of his neck was when he was proceeding on leave and handed over the charge of the RTC to the officiating incumbent. Clearly, he trusted himself more than he did the Quarter Guard. It was this 'Sasta aur Tikau', simple and reliable approach that brought him so far.

The Officers' Mess

Before I narrate some incidents of my unit officers' mess, I wish to apprise the readers with the concept of an officers' mess; something more than what meets the eye.

The officers' mess is the home of bachelor officers in the unit. They live there, dine there and spend all their private time there. Well, of all the things, dining in the mess is of the least significance. The officers' mess is one of the two most important institutions in a unit; the other being the unit quarter guard.

The officers' mess is the repository of the unit history, customs and traditions. It is here that the entire history of the unit is displayed in the form of war trophies, gallantry awards, silver trophies etc. A gallery of the war heroes and battle history tells the tale of the blood, guts and glory of the unit, recognized in the form of battle honours. A great grand son of an officer can walk in to a unit officers' mess, and he will be told more than what he knows about his grandfather, well archived. The units maintain an archive which contains all the details of its history and preserve all articles which have some historical significance such as the first type writer, the first unit routine order and a record of all the firsts etc. It is here that the unit forges the battle winning camaraderie and brotherhood that makes them a band of brothers. The esprit de corps developed under the watchful gaze of their war heroes, peeping from the picture frames, and under the shadow of war trophies makes them a proud unit that will prefer death to defeat. It is here that unit traditions are forged.

The young officers, on joining the unit, spend their first few months learning unit history and history of every single item in the mess. Officers are required to pass the regimental history test periodically.

The soldiers too visit the mess occasionally and 'feel' the regimental history which becomes the foundation of their regimental pride and motivation.

Well, to illustrate the point in brief, 4 GUARDS (1 RAJPUT) has a centre silver piece in its mess. It contains shreds of the Colours called 'Lake and Victory', presented to the unit by Lord Lake in 1803 for sustained valour and sacrifice. The unit draws its inspiration even today from this symbol of valour and regimental honour. The episode is indeed militarily awesome and worth knowing.

In February 1805, the battalion launched a series of attacks on Bharatpore fort. The last attack was beaten back just as this Colour, by now bullet ridden and in tatters, was being planted on the ramparts of the fort. The attack had to be called off as the battalion suffered two hundred dead in a matter of few hours. It was decided to replace the tattered Colours with new ones. On the day of the ceremony, the tattered Colours had mysteriously disappeared from the unit quarter guard, never to be found, nor pursued.

25 years later the battalion attacked and captured the same fort and planted their new Colours. The shreds of the tattered Colours which had mysteriously disappeared from the quarter guard 25 years earlier, had reappeared, proudly fluttering in small pieces alongside the new Colours, precisely from where it was beaten back 25 years earlier. The forefathers, who were stung by their defeat had passed on the shreds of the Colours to their children to restore the unit's honour some day; and they did. [*Source: A Matter of Honour, by Philip Mason, Regimental history of 4 GUARDS (1 RAJPUT)*]

Every recruit takes oath to lay down his life for the honour of the paltan (colloquial expression for unit/battalion) on these two hundred years old shreds of cloth; a symbol of their honour.

All units have such trophies which are symbol of their

pride. Young officers are groomed under such fine traditions in the unit officers' mess. That is what the officers' mess is all about- the crucible of regimentation and brotherhood.

The units are authorized many tradesmen including house keepers, hair dressers, equipment repairers, chefs etc. They are steeped in regimental spirit and are an indispensable lot without whom no unit can function. Officers' messes too are authorized mess staff i.e. stewards, chefs and house keepers etc. While their primary job is administrative, they are soldiers first and cook or a waiter later. They do all that a combat soldier does; endurance training, battle craft, drill and shooting etc. They take great pride in being part of the unit and keeping high standards in the mess, and often tell stories of what used to happen in the old days and how things have deteriorated over the years. Many of them go on to become institutions by themselves, and will not allow anyone to play the fool with mess decorum and customs. Many have made the supreme sacrifice in battle and have been decorated for valour. Strict mess decorum is maintained in terms of conduct, dress regulations, timings etc. Periodical dinner nights and guest nights maintain the formal decorum, train the young officers and the ladies in mess etiquettes, and the mess staff and the regimental bands in mess drills and customs.

The Wonderful Tradesmen

Second Lieutenant Robin Cherian was recently commissioned in 5 GUARDS, then in Siliguri. On a Saturday evening he walked into the mess. Seeing no one, he shouted "Koi Hai." No one responded. He moved deeper and located and an old timer waiter, an institution in himself, a happy go lucky fellow called Havaldar Hazari. Seeing no one around, Robin asked Hazari.

Robin : *"Saab log kahan hai?"* (Where are the other officers?)

Waiter : *"Bahar gai hain, bola aaj khana bahar hee khayenge,"* (They have gone out, said they will eat out today). Robin was much relieved deep down, for dinner in the mess was invariably a tough experience for a subaltern with all the grilling.

Robin : *"Khana bana hai ki nahi?"* (Has the dinner been cooked or not?)

Waiter : *"Nahi bana hai saab."* (No it has not been cooked).

Robin, saying *"Khana kyon nahi bana hai?"* (Why has the dinner not been cooked?) entered the cook house and saw that there was some dal and vegetable in two dongas. Pointing at them, he said, *"Yeh kya hai phir?"*(What is this then?) The waiter replied with a squirm *"Yeh koi GUARDS afsar ke laik khana hai saab. Aap bolo toh de deta hoon."* (This is no food for a GUARDS officer. If you tell me I'll serve it to you).

So that was the spirit of the mess waiter, who believed in high standards for a Guards officer.

Yes Sir! Yes Sir! Three Glasses Full

Havaldar Meherban was a mess waiter in 5 GUARDS till the mid eighties. He was a roly-poly, fair looking Garhwali with large eyes and hanging cheeks. No taller than the shortest permissible height, he was as broad as he was tall and sported a thin moustache. His trade- mess waiter, had done his health a lot of good.

Befittingly, he was called 'Napoleon'. Not seen unless called thrice, and almost never heard, he conveyed an impression of sincerity, solitude and contentment. Caught unawares, he could be seen threatening the Mess Havaldar.

"If you trouble me too much, I will quietly throw away one silver piece of the mess, and you'll keep paying all your life."Looks can indeed be deceptive.

We had very few dining members in Siliguri in 1979. Major Harpal Thind, a fitness freak, would take a glass of milk after dinner every day.

One Saturday evening some married officers joined in for a meal in the mess. When the dinner was over, as usual, Meharban brought a glass of milk for Harpal Thind. Harpal asked him to bring milk for other officers also. Meharban stood stone faced looking at Harpal. Since no one declined the offer of milk, he picked up the glass and disappeared into the kitchen. A little later, he reappeared with four glasses of hot milk on a tray. Every one enjoyed the good milk.

Harpal then asked Napoleon, "It is so hot and the fridge is not working. How much milk do you keep in the mess, that you suddenly produced five glasses?"

"Only one glass of milk for you saab. No one else drinks milk after dinner and then milk is brought in the morning," replied Napoleon innocently.

"How did you arrange so much milk now?" asked Harpal.

"You asked me to bring milk for everyone. There was no

more milk in the mess, I had to do something. So I added four glasses of hot water and sugar," replied Napoleon. The innocent face with which he could survive blue murder, had passed another test. There is not to question why, a task accomplished as ordered.

Napoleon at his Imaginative Best

Napoleon was never out of news for long. He earned the wrath of the unit's first lady once. The commanding officer walked into the mess with his wife one evening. Later we sat down for dinner. One officer observed that the commanding officer's wife was waiting for long for chapatti. Napoleon would come out of the kitchen with two chapattis at a time and give it to one or the other officer, bypassing the lady every time.

Next time, when he came from the kitchen with chapattis, the officer asked him to show a chapatti to the commanding officer's wife *(Napoleon Madame Saab ko chapatti dikhao).* Napoleon obediently walked straight to her, held the plate on left hand, picked a chapatti with his right, hung it in front of her face, turned it around a few times, placed the chapatti back on his plate, walked across and gave it to another officer. As ordered, he had shown the chapatti to the commanding officer's wife.

What would you do, if you found yourself in a situation like this? The mess secretary would count his days and the second-in-command, tear his hair. The rest would of course laugh, as we all did.

Lesson learnt, be precise with your instructions to the men, leave no scope for interpretation at their level.

I am Dining out Today

The officers' mess in units is an important regimental institute. There is a mess committee comprising number of members to run the mess efficiently. One of these is the food member who is responsible for ensuring high standard of food and all aspects related to it. In the early seventies, we had in our unit a very fine regimental officer by the name of Lieutenant Prashar. Prashar was a bit on the heavier side and was famous for his liking for good food, and lots of it. For this trait he was appointed the food member. He did a great job and all dining members were very happy with the food. One day the officers found the food to be particularly routine, some lousy vegetable and dal. Everyone was surprised at the lousy food and noticed that friend Prashar was missing. That was uncharacteristic of Prashar, everyone thought.

When asked, why the food was so awful, the Mess Havaldar replied, *"Saab aaj ka menu bhi Prashar saab se hee poocha tha. Bole aaj mein dine out kar raha hoon, toh aaj total ka bana dena."* (Today also I asked Lieutenant Prashar for the menu, he said he was dining out today so make some damn crap).

Koi Haip

In the battalion we would pay particular attention to the selection of our Regimental Police Section and the Mess staff, for they were on the show window and their looks spoke louder than words.

In the snow covered Sonamarg in Kashmir, we were doing winter training in 1985. The officers' mess was in a snow tent. We had invited the brigade headquarters staff for lunch. The officers and men wore special clothing blue wind cheater over their uniforms which concealed their badges of rank. One officer from the brigade headquarters entered the mess and found a smart young man standing at the entrance to the mess. "Hello, I am Captain Mitra, the Staff Captain, we have not met before, have we young man?" he said stylishly.

"Ram Ram saab, Guardsman Rajkhowa, mess waiter," replied the young smart Guardsman. Mitra looked around to see if anyone had noticed. Some had.

As the lunch proceeded, one of our officers looking for a waiter shouted "Koi Haip," lo and behold a Guardsman quickly responded. After a while another officer wanting to call a mess waiter shouted "Koi Haip" and another waiter appeared. It happened the third time and third Guardsman appeared. Mitra looked puzzled and asked one of our officers, "Amazing, you have so many jawans by the same name!" He sat quite sheepish when he was explained that the officer was only shouting *"Koi Hai"* (Any one around?) and any mess staff who hears comes over. *"Koi Hai"* when shouted sounds like *"koaip"*, difficult to understand. Since our men were from the North-East, it was often misunderstood by the visitors as their name.

Mitra went through the rest of the afternoon abiding by the old adage - "Silence is golden".

Qayoom's Dilemma

My unit was posted in Poonch in 1991. The unit rear was in Poonch base and the battalion headquarter was located well forward, over an hour's drive on a rugged bumpy road. We had a youngster, Second Lieutenant Narinjan Singh alias Qayoom Khan, a slightly built specky little surd with a good zestful attitude to life. Doing the battalion quartermaster (Logistics officer), he was located at the battalion rear. That is an important but a thankless little job. Units never notice the quarter master if all is going well. It's only when things go wrong that he is remembered and put on the chopping block.

Young in service, Qayoom would err once in a while. An observation or a reminder from the brigade headquarter would bring upon him the commanding officer's (CO) wrath which was fairly frequent. It was very rare that this fellow ever came out of the battalion commander's office without a mouthful. As a result, he had started avoiding the 'speak' cases. Every time the CO called him in connection with a letter, he would say "I've spoken sir". In effect, in every speak case with potential for trouble, he would simply write 'spoken' next to where the CO had written 'speak'. The CO being a busy man was able to see the quartermaster very rarely, and he would get the benefit of doubt and get away with it.

Having completed two years service in field, Qayoom became due for his Captain rank and was to be promoted that day by the CO. I was doing the adjutant at the time. He was called to the battalion headquarter for the pipping ceremony by the CO. This is mostly done in the CO's office or sometimes in the mess. The ranks are soaked in a glass full of liquor which the officer empties in one swig. I congratulated him on his promotion to be. He said,

"Sir, congratulations baad mein dena. Aaj toh speak cases bhi bahut hain, bahut joote parenge. Ap hee batao pahle speak cases karoon ya pahle Captain banoon." (Congratulate me later. There are many 'speak' cases today. It does not seem to be a promising day. There will be boot parade for sure. Please tell me whether I should get over with the speak cases first or get promoted first and then go for the 'speak' cases).

I said, "Don't worry about the 'speak' cases, we will first make you a Captain."

He said with his mischievous smile, *"Nahi sir, pahle speak cases khatam kar leta hoon nahi toh wahi hoga, 'sar mudate ole pare', Captain banate hee joote parenge. Aur phir poore Captain rank ke period men joote hee parate rahenge. Aur Captain rank chalta bhee dus saal hai."*

(No sir, I think I will finish the speak cases first, and get promoted later. Otherwise, I will get a boot immediately on promotion and continue to get kicked throughout this rank. And you know Captain rank lasts over ten years, far too long to take such a risk).

I guess he was right. He did the 'speak' cases and as he had appreciated, got kicked no end and then got promoted to Captain. The CO told him "I am promoting you even though you do not deserve this rank yet. I expect better results." Not a very encouraging way to be promoted. He put on the rank and enjoyed a peaceful twenty four hours before he was back again for the next 'speak' case and the bollicking from the CO.

Over a period of time Qayoom had got on the wrong side of the CO. One day he walked into my tent after his 'speak' cases meeting with the CO. "How was the day Qayoom?" I asked. He said with his usual jest, "Nothing new, the boot parade continues as Captain."

I was sharing the tent with Major Manjit, an Artillery OP officer who used contact lenses. Manjit sat on the bed with a bowl of water removing his contact lenses after the day's work. Qayoom saw something in the bowl and looking at Manjit's face said:

"Wait one, funny sir, your one eye is black and the other is brown. How come?"

"Well, I use contact lenses. I chose brown contact lenses, just for fun," the officer replied.

"You *chose* brown colour! You mean they are available in all colours," asked Qayoom.

"Yes."

"Even blue colour?"

"Yes , even blue colour," replied Major Manjit.

"Wow, that is great. I think I will also switch over from glasses to blue contact lenses. The way things are between me and the CO, it seems to be the only way of becoming CO's blue eyed boy."

Qayoom's zest for life never diminished no matter what the circumstances. He had earned this alias name later in the valley when he temporarily assumed Islam and did commendably well in the counter terrorist operations there. He later resigned his commission. The Paltan misses this wonderful guy.

Let the Sleeping Lamb Sleep

The battalion was doing an exercise in the deserts. We had been on the move for almost four days with little rest or sleep. All movement was by night. The sweltering heat of Rajasthan allowed you no sleep or rest by the day. Lieutenant Colonel Gill, the second-in-command was travelling in a rickety Jonga. The driver, Guardsman Duty Ram Deka, also much short of sleep, kept driving like a robot with a fixed gaze, eyes wide open, rarely blinking fearing he may fall asleep. Gill, who was himself dozing, noticed the driver was finding it hard to keep awake at the wheel.

Gill asked the driver to stop and took over the wheel. They must have been driving for a while, when, suddenly suffering severe bumps and jolts, Gill and the gang were shaken up awake. He found himself in the middle of a field many yards off the road, Jonga nearly rolling over.

"What happened?" asked Gill.

"We are in the chana field," replied Deka.

"I know that you idiot, but how did we get here?"

"Well, saab you went to sleep."

"When?"

"Five minutes ago," replied Deka.

"And what has been happening since then?"

"I have been driving from the co - driver's seat," replied Duty Ram Deka dutifully.

"Why did you not wake me up?"

"Sir, you were too tired. I thought let saab take rest for a while, till then I will drive from the other side. It was all fine till a sharp curve came up, then we landed in the field, and it was too late to wake you up saab, you had woken up yourself," replied Deka, with utmost sincerity.

Foolish or loyalty - your call.

All Pups and Dogs will be Shot

Major Pup Mann was doing the second -in - command of 4 GUARDS (1 RAJPUT). The dogs were a big menace and the adjutant decided to sort out the problem for good. He sent out a circular saying that by the next Sunday all pups and dogs in the unit area will be shot dead. Keep yours tied in.

Pup Mann promptly wrote on the circular, "All but me please."

Colonel Pup Mann had an excellent way of conveying his mind without directly saying it. During the collective training in Sikkim in December, the battalion was camping in the tents. We were informed that the corps commander, the divisional commander and the brigade commander would be coming to the unit the next day and shall stay overnight.

So many top brasses are a handful even in the best of times.

In those circumstances, it was blue murder. Unperturbed, he called a small conference and conveyed what he wanted done. Then he said,

"Gentlemen, it is a pleasure to have such senior officers staying with us in the battalion. Don't worry about the tents, we will give them the best we have. I know you will look after the corps commander, for he is a big man. I know you will also look after the divisional commander, he too is a big man. Don't forget your commander, remember, a dog knows his master, not his master's master." Well said.

Pup Mann retired as Lieutenant General after an illustrious career. Good regimental officer that he had been, he married his daughter to an officer in 5 Guards (the battalion he commanded) and sent his son to 4 Guards (the battalion he was commissioned in to). Over the years, he had collected three cars. One, the latest, was the diesel Tata Indigo that he regularly used. The other two were old models Fiat and 118 NE, both obsolete and out of production. They were mostly run to charge the battery and little else. His driver was a faithful retired soldier who had served with him for long during his service and chose to stay with him after he retired.

His son, Pips Mann, a fine young officer in 4 GUARDS would proudly say, "Dad, we are a three car family." He was put in perspective when General Mann's faithful soldier suggested, *"Saab in doh car mein se ek ko bech ke tussi mobile le lo."* (Why don't you sell one of these two cars and buy a mobile phone instead). So that was what he thought of some of the cars of this three car family.

PART V

MILITARY SERVICE

Through the Service

In the earlier part of the book we went through some wonderful experiences from the National Defence Academy, Indian Military Academy and my experiences during the regimental life. I shall now take you through some exciting exchanges from different aspects of service. During the course of more than three decades of service, I have met many colourful characters, both officers and men, who have added zing to life. These are experiences of many others, as well as my own, and stories that have been around in the army folklore since long.

Any army, as they say, is as good as its officers, for men always deliver. That's why the saying, *"It is better to have an army of deer led by a lion, than an army of lions led by a deer."*

Some General is known to have divided officers in the following four categories, and truly, you do find them everywhere.

- *First - Intelligent and hard working.* They make excellent staff officers.
- *Second - Intelligent and shammers* (military term for shirkers). They make excellent commanders.
- *Third - Foolish and shammers.* They are fairly harmless. Not much problem, the army can manage them.
- *Fourth - Foolish and hard working.* No organization can afford to keep them except at its own peril, for they are a danger to own organization.

I had an experience of the last category. A very hard working and a sincere officer was part of my team when I was commanding my battalion. Give him a task and he would go to ridiculous details. The units those days got a lot of their stationery from the military stationery depots. Barring a few forms, the quality of this stationery was rather poor and outdated and no one ever used them. Often they were just dumped and left to their fate, and occasionally used for eating samosas or cleaning the tables. The clerks preferred to type out the forms rather than use these outdated ones.

This officer was detailed to take stock of the stationery. I was told that he had been doing the stock taking for days and not issuing anything which had affected the work in the office. I was quite amused to learn that he was taking his job of counting the useless stationery stupidly seriously. Counting every single page of the useless dump, some of it already in yellow tatters and falling apart. And above all, not prepared to issue anything till the whole dump was counted.

No not Now

Cadet Neeraj Bali and myself were in Lima Squadron in NDA. He was a course senior. Neeraj was a gentleman, an intelligent bloke with good communication skills and a good sense of humour. He did well at everything except that mattered most in NDA, the outdoors. Outdoors were not his preferred cup of tea. Not blessed with the agility and figure of Nadia Comaneci, he struggled to get through the PT tests. Hand spring and upper circle were the worst things ever invented by a maverick, he believed, and some sadist clot included it in the PT tests in NDA. During games he preferred to keep the score, or at worst be the referee. He owned a squash racket which he brandished in front of first termers who had never seen a Squash court. Inter squadron cross country competition, he believed, was violation of human rights and cruelty to cadets.

Well, in NDA, you are a good cadet if you are good in outdoors, otherwise, you are useless. A six star academic torch has no meaning if you cannot play foot ball for the squadron. So, few ever noticed his dramatics and debating skills. I was reasonable at outdoors and thus considered one eyed amongst the blind.

Normally, first termers and second termers are like a horse and the grass. They can't be friends. But we got along well and became friends. As a second termer, Bali was gentleman enough to offer me the privilege of addressing him by name, rather than 'sirring' him. I said, how does it matter? We are friends, you are senior, so I shall continue to sir you and so it stayed.

With passage of time, we crossed the phase where physical fitness alone was the yardstick of competence. Bali proved to be from the first category and had started to make his presence felt, was decorated for gallantry and nominated for the Higher Command Course, an indicator of a bright future.

In 2000, the Higher Command course was visiting Jammu where I was commanding my battalion. Being around my seniority, majority of these officers were known to me and I invited them for cocktails in our mess. Everyone was enjoying the evening with our wonderful Pipe and String band music. I was chatting with Neeraj Bali and sharing jokes of days bygone including the NDA days. And then I reminded him.

"Do you remember, in NDA, you had offered me the privilege of calling you by name? I had declined saying it does not matter and I continued to address you as sir, despite being good friends."

"Yes, I do remember after all these years," he said.

"I did not exercise the privilege then, can I do that now?" I asked.

"No, not now. Now I am doing well," said Bali.

Well, time passed and he became a Brigadier. As for me, I missed the board to Brigadier the next year. Bali called to express his condolences as if I was already in the grave. Next year, the army surprised me by approving me for promotion. Bali again called to compliment me for my resurrection from the grave.

I said, "Well, can I call you by name?"

He asked me, "Why this now?"

I replied, "Because, now I am also doing well."

Hats Off to India

It was on 26/11 that Pakistan based terrorists attacked Mumbai. The soldiers of the NSG brought the whole episode to an end after a gruelling battle, making huge sacrifices in the process. When over, the soldiers were transported back to barracks in rickety state buses ready to fall apart. The issue caught attention of many and was highlighted by the media. They compared this with the Indian cricketers, whose contribution to the society, be what it is, are transported from one place to another in expensive luxury buses. Shameful as it is, it was flashed by media under the heading 'Hats Off to India'. I got this article from a friend on the net and forwarded it to then Brigadier Bali, commanding a brigade. Humble man that he is, I promptly got a reply: "Hats off to India, they have selected me to head the Indian training team in Lesotho."

A Rare Privilege

The 'Georgette patches' are the red collar dogs worn on the collars by officers of the rank of Colonel and above in the Indian Army. In fact, they are worn by armies all over the world. In our army, they are referred to as 'red tabs'.

Generally speaking, they are associated with stiff necks and high nose, not always popular among juniors.

In the Army Medical Corps, the doctors get a few years ante date seniority on commission. There was an Infantry officer married to a lady medical officer. She was promoted to the rank of Colonel with her ante date seniority, while he was still a Lieutenant Colonel. The lady wore red tabs while the husband did not.

During a party, one officer asked him, "How do you feel, being bossed around by seniors in the office and having to put up with a Colonel wife at home and saluting her."

"Good," he replied. "In the army, I am the only junior who has the privilege to have a red tab in bed."

The Great Army Drivers

A military driver can be a bit of a nuisance to traffic on the road. That is because more often than not, the areas where he drives military vehicles for better part of his service, he is the lone ranger on the dusty isolated roads in some remote part of the country. There, being the emperor of all he surveys, he has the freedom to set his own traffic rules. Quite often, he is handed over the vehicle before he is adequately baked as a driver. If he lands up on the Delhi roads, he is like a cat among pigeons and can be a handful creating enough confusion, sending everyone on the road scurrying for cover. Driving skills apart, his ability to fix a mechanical problem is next to nonexistent.

Vinod Pillai from 12 GUARDS was travelling along a bumpy road in a jonga. The vehicle stuttered, coughed a few times and came to a grinding halt. The driver got off the vehicle, opened the bonnet and tried his limited skills to fix the problem, but to no avail. He searched around, grabbed a stone and perched himself up on top of the vehicle. Opening up the air filter, he hit the stone all over the carburettor, hoping like hell that the trick would work.

The driver returned to the wheel and tried to start the vehicle without success. Vinod, who all this while was watching patiently felt the vehicle may have run out of fuel. He asked the driver in Hindi if the vehicle had enough fuel. "Very much," replied the driver in English with confidence. When, in spite of all his efforts the vehicle did not start, Vinod again asked the driver if there was petrol in the vehicle. The driver again replied *"Han saab, bahut hai."* (Yes sir, more than enough).

The officer then settled down again hoping the driver will fix the problem. When nothing happened despite the punishment being meted out to the carburettor, and by this time worried that he might break the damn thing, Vinod walked around the vehicle and peeped at the fuel gauge. Not particularly surprised, he saw the fuel gauge needle pointing at `E`. He then called the driver and said,

"Mein aap se kitni bar poonch chukka hoon, ki petrol hai ki nahin, aur aap bol raha hai ki bahut hai saab. Yeh dekho kya hai," said Vinod pointing at the fuel gauge. (I have asked you so many times if there was fuel in the vehicle or not and you have been telling me, yes there is enough, look at this fuel gauge.)

Driver : (Peeping at the fuel gauge) *"Theek toh hai saab."* (It's perfectly all right sir).

Pillai : *"Kya theek hai, needle kidhar hai?"* (What is right, where is the needle?)

Driver : *"Saab 'E' par hai."* (Sir it is on 'E').

Pillai : *"Toh 'E' ka kya matlab?"* (So what does E mean?)

Driver : *"Saab 'E' ka matlab* 'ENOUGH'." (E means ENOUGH).

Pillai : *"Aur 'F' ka matlab?"* asked Pillay (And what does F mean?)

Driver : "FINISH," replied the driver poker faced.

Well, the army vehicle numbers include a prominent arrow pointing upwards, called broad arrow. This is to ensure that the driver knows which is the right side up – if the vehicle topples over!

Our division was doing winter exercise in the Rajasthan deserts. The Colonel General Staff (Colonel GS) was a Sikh officer. His driver was another surd from Punjab regiment. The desert roads are straight roads in a desolate landscape with little or no vegetation that enables you to see miles. Accidents are thus rare. Even if the vehicles go off the road, there is no harm done.

The Colonel GS got a report that his driver had met with an accident and had killed a camel. The poor villager was devastated as the camel was his only source of livelihood. He wondered how his driver had managed to bang into such a huge thing as a camel. He called his surd driver, bullshitted him and said, "You idiot, how did you hit such a big thing as a camel in a desert where you can see miles. Are you blind? You must have been over speeding."

The driver replied *"Saabji, mein toh dheere hee ja raha seega, who toh oont vadi speed se a raha tha saabji.* (Saab, I was rather slow, it was the camel which was over speeding).

A battalion with Sikh troops was recently posted to a peace station from field. Less than a week later, two of their vehicles collided outside their main gate. As is customary, a court of inquiry was ordered. The statements of the two drivers were as given below.

Sepoy Kulwinder *"Saabji, yeh jo satinder hai na saab, yeh bahut careless driver hai saab. Bahut tej gaddi chalata hai. Sari paltan ko maloom hai. Jaise hee mein main gate se bahar nikala, maine dekha kee yeh satinder badi teji se gaddi chalata hua main gate kee taraf aa raha hai. Mujhe paka maloom tha yeh jarur rora pane wala hai. Maine jaldi se apni gaddi rok kar khari kar li. Is ne seedha a kar meri gaddi mein takkar mari saab. Mein toh ekdam khara tha saab, saari galti is kee hai."*

(Saab, this Satinder is a very careless driver. The whole battalion knows he drives very fast. The moment I came out of the unit main gate, I saw him speeding straight at me. I knew he is going to create some problem, so I stopped my vehicle and stood at one side. He came and banged into me. The entire fault is his).

Statement of Sepoy Satinder was exactly the same.

The presiding officer of the court of inquiry was in a quandary. He followed the principle that in case of doubt screw both, and he did.

By 'D- Day' Sir...

- *D-Day : D-Day is the pre-decided day for commencing an operation.*

As a legacy of the British army, English language has come to stay as the working language amongst the officers in the Indian armed forces. It is embedded so strongly that one has to make a deliberate effort to converse in Hindi. And when you do so, before you realize, you tend to switch over back to English again. The troops therefore associate officers as being English speaking.

In the army there is a type of commission called Regimental Commission (RCO). They are the outstanding Junior Commissioned Officers (JCOs), selected for commission as officers. When they are finally selected, they overnight change from Hindi speaking JCOs to English speaking potential officers. The sudden change to English is not so smooth, when one has not had formal education in the language. In an effort to prepare themselves for what lies ahead, some insist on speaking in English, even if it means confusing the hell out of everybody around.

One Gurkha JCO who was selected for commission was tasked to construct a Divisional Battle School. The brigade commander paid a visit to the battle school to see the progress. He met the JCO and spoke to him.

Commander: *"Saab, kaam kaisa chal raha hai?"* (Saab, how is the work progressing?)

JCO: "Saab, going very well sir."

Commander: *"Saab kya time per battle school tayyar ho jaega?"* (Will the battle school be ready on time?)

JCO: "Yes saab, definitely."

Commander: *"Takriban kab tak tayyar ho jaega?"* (When do you think the school will be ready?)

JCO: *"Saab by D-Day."* (It will be ready by D-Day).
Commander: *"Aur D-Day kab hai saab?"* (And when is the D-Day?)
JCO: *"Jab aap bolo saab."* (Whenever you say sir).

Always an Occasion to Beat it up

A Corps headquarters is a top heavy organization, with large number of senior officers. The Corps officers' mess hosts social functions pretty frequently. These officers' messes have one officer who is detailed as the food member. The food member is responsible for all aspects of food and upkeep of the mess. All such unpopular duties fall in the lap of junior most officers, invariably the Captains in the Corps headquarters. He is required to be present at the formal functions and ensure that all goes well, strictly meaning nothing must go wrong. Essentially overseeing the Mess Havaldar, it is a thankless job, none relishes doing it and would dodge it if they could help. In the units, officers' mess is a sacred institution and well looked after, in higher headquarters, it is no one's baby.

One day as I walked in to a Corps officers' mess, I saw two young Captains drinking away to glory. When asked, what was the occasion for this beat up? One said, "I am celebrating, I have been relieved of the duties of food member of this mess." And you, I asked the second. "I am drowning my sorrows, I have been appointed one instead," he replied philosophically doing bottoms up and ordering another large drink.

Army and Horticulture

The army has done a yeoman service in preserving the environment. It is my humble guess that the army alone has planted ten times more trees than the forest department, seeking no credit for it. We now have Territorial Army battalions (Ecological) which are changing the environment and landscapes by revitalizing lost lands through tree plantations and conservation of forests. We often joke that every commander's fad is to plant trees whether there is space or not. Bhatinda and Bikaner cantonments have been converted from semi deserts to oasis. The cold whiff of air that hits you in the face as you drive into these cantonments is unmistakable and carries the reassuring message of army's contribution in this field. The cantonments are the preferred walking joints of civilians. They drive miles to take a walk in cantonments whether it is Ambala or Jammu.

Well, posted in Khreuh near Srinagar in the mid eighties, we had a commander who was a tree planting buff. Instructions were passed that all units will plant a thousand poplar trees. There was no place for such large number of trees in the small battalion area. Nevertheless, many trees were planted. A few months later, the commander wanted to see the progress. That came as an unpleasant surprise. On the given day our commanding officer, Colonel Ranjit waited to receive the commander at a vantage point that gave a good view of the battalion area.

The commander came and surveyed our plants that had been planted along the road.

"Good job, Ranjit. These plants look good and healthy." Then as he turned to walk away, he slipped and reached for the six feet tall plant to steady himself. And lo and behold, what happens. The plant comes out in his hand, albeit with no roots and a sharply cut stem. He looked with disbelief at the healthy poplar plant without roots and looked at the commanding officer who was equally surprised.

Well, the unit had not planted trees there. When the unwelcome news of the commander's visit came, contingency plan was put in place. Six feet long poplar branches were cut and planted the same afternoon.

This is a 'fire fighting technique' often employed by the units. They will create a rose garden overnight in deserts which will disappear even faster.

We were posted in Bhatinda along with a Gurkha battalion. The corps commander also was Gurkha officer. The Gurkha commanding officer was asked to meet the Gurkha General in MES Inspection Bungalow. He was told that the corps conference was scheduled in mid February and the army commander would be staying there.

He said, "I want this whole garden in full bloom when the army commander stays here ."

"Sir, but the flowers don't bloom in February," the colonel said.

"You think I don't know," said the formation commander, "that is why you are here."

The Gurkha commanding officer understood the Gurkha sentiments and said, *"Honcha huzoor."* (Will be done, my lord).

The task was given and in the true traditions of Gurkhas who can do anything, the task was addressed in full earnestness. The formation commander inspected the garden in the first week of February and was satisfied with the progress. A few days before the conference, the formation commander called the commanding officer to his office and said, "You know the conference has been postponed by four weeks and he would now like the flowers to bloom in mid March. *"Honcha huzoor,"* was the smart reply, not that he had an option. And, lo and behold, the same flowers that were all set to bloom in mid February, bloomed in full glory, this time in mid March when the army commander arrived. How they managed, only the Gurkhas know and they will not share their secret.

The Bear Trap

A few decades ago, there was a three star General. His children were grown up. His son had settled in USA and had married an American. Visiting India for the first time, he wanted her to see the army style. A hunting buff himself, he decided to show her a bear hunt in a real environment. That was the time when the embargo on hunting these animals was still not in place.

The General said that thirty years ago when he was a young officer, there was a beautiful forest in terrai region of Kumaon. They would go game hunting and have good fun. He asked the divisional commander if he could organize a hunt when his 'foren' daughter in law comes. The task was delegated from the division to the brigade, and as always, it landed in the reassuring lap of an infantry battalion.

There were many issues that comprised successful accomplishment of this task. The first was to carry out reconnaissance of the target area which the General had seen in a different era. The commanding officer not wanting to leave anything to chance carried out the reconnaissance himself. To his horror, the jungle in question was nowhere to be seen. The forest had become a casualty of ever increasing population pressure. The once upon a time forest was a flourishing residential colony. But who would bell the cat and tell the boss. No one. Some solution had to be found.

After a lot of petrol had gone up in smoke, some wooded area no more than three hundred by two hundred metres was found somewhere. It was decided to bring him on a pitch - dark night and take him around in circles to simulate a long drive in the forest. There were no animals worth the name except stray dogs and wandering cows. They were hardly animals to be hunted. The next problem was to find a suitable animal. Since the General had mentioned bear, all efforts were focused on managing a bear. Bear is not a commodity that can be taken out of a hat. Finally, after much brain storming, they decided to buy one from bear owners who use bear dancing to earn a living. Number of search parties were launched in different directions. Many bear owners were contacted who refused to sell their bears since their livelihood depended on it. All incentives and offers failed to persuade any one to sell one. Finally, an arrangement was worked out with a bear owner to loan his bear for a night, at a large sum, on the condition that nothing should happen to his bear. That was a bridge they would cross when they come to it. For the present, getting a bear was the only thing of interest. The hunt was organized and fixed for the weekend after a thorough study of moon timings to ensure there was no moonlight. The route to simulate a long drive was selected and gone over. A splendid machhan (platform on a tree) was made with not too a cumbersome ladder so that the 'foren' daughter in law could climb without much problems. A supply (ration) goat with one hoof already in the grave was tied to the tree to simulate a bait.

The cavalcade of the General slowly drove in pitch-dark without head lights. Everyone maintained pin drop silence. The American was truly excited at the thought of a live hunt -Maharaja style, about which she had only heard. The vehicles were taken away by the drivers as the VIP delegation climbed the machhan one at a time.

The bear party consisted of the bear, the handler who was less than fourteen and a 'one size fit all' trouble shooter Second Lieutenant. As practiced, the bear party was hidden in a dense bush in a three men trench.

The VIP was told that if they were lucky they could even expect a date with a leopard. As they waited, the supply goat repeatedly tugged at the rope, which, true to its military heritage, gave way after a while. The goat went around the tree and started walking away. Someone had to go down and tie the goat again. There was no volunteer to do this task, clearly beyond the call of duty in peace time. Well, finally, as it often happens, it became the privilege of the junior and it fell upon the poor ADC to rise to the occasion. Having no options, he went down like lightening and climbed the ladder faster than the fittest monkey.

The group waited patiently on the machaan as they were served tea and snacks. Suddenly the commanding officer put his finger on the lips and drew their attention to some sound as the supply goat became restless.

At the pre-decided time the Second Lieutenant, the commander of the bear party, pressed the arm of the bear handler. The handler tapped the bear on his butt with a stick. The bear then climbed out of the trench and brushed a bush shaking it wildly. The General and his son raised their double barrel guns. Then there was a deafening silence. The bear emerged from the bush, stood on hind legs and started dancing in circles shaking his head from side to side. The General lowered his gun and said, "Wait, there is something amiss. Never seen a bear dance in the jungle at midnight."

The whole game plan was exposed. The bear handler had made a mistake. In that darkness, he tapped the bear on the wrong butt. In the normal course that was a signal for him to start dancing. All the commanders from the battalion commander upwards knew their careers had come crashing down with the tap on the butt of the bear. The Second Lieutenant was the convenient scapegoat for lack of coordination. Fortunately, rules do not permit reducing a Second Lieutenant to the lower rank.

The General said nothing, smiled and climbed down the machhan with his American daughter in law in tow. She did not see a bear hunt. She saw something better, a dancing bear at mid night.

Well, I cannot be sure of the veracity of this story, but had heard this a long time ago. One divisional commander was fond of angling. A picnic was organized by the brigade near a lake in East Sikkim, famous for trout fish. The divisional commander was always the first to catch fish. How they managed that, only they knew. Everyone was happy. The General for his angling skills and the brigade staff for making their own day. Well, I am not sure, the story goes that there was apparently a subaltern who volunteered to swim under water to hook a live fish to the General's tackle. Mercifully – I am told they did not try that.

Running for the Sword

In the NDA we had a short statured course mate by the name Ruperi. Some affectionately called him chotu. He was a hilarious, happy go lucky character hailing from Bhopal and spoke with a typical Bhopali accent. He came to be known as 'Soorma Bhopali.'

His only claim to fame, which he claimed and announced every week, was that the famous film actress Jaya Bhaduri and he studied in the same school. If I remember right, it was Campion School, Bhopal. He was a naval cadet who joined the Naval Academy after the NDA. In all service academies the best all round cadet is awarded the 'Sword of Honour' which is a unique honour. The recipient of the 'Sword of Honour' of the course at IMA and friend Ruperi were from Lima squadron and were good friends.

On having heard of the achievements of his friend, Ruperi could not help broadcasting his own. He wrote to his friend:

"Partner, my heartiest congratulations on your getting the Sword. I am proud of you. You will be surprised to know, that here in the Naval Academy, I was also in the run for the 'Sword of Honour', but unfortunately someone else got it and I kept on running."

A Matter of Perception

- *OC: Officer Commanding a unit*

I have had a reasonably good record of service all along. The expectations from colleagues were high. As time flew by, we had reached the stage for promotion to Brigadier. The promotion board took place, and I never tried to find out the board result for I knew someone would give me a call either way. A course mate called me up saying that he had heard I had missed the board and that everyone was shocked. I said I had never bothered to find out, but if this is what he had heard, then it must be true. I then started laughing. He asked me, "This is not good news, what is there to laugh?" I said, yes, there indeed is something to laugh about and narrated this story:

I was commanding my battalion in Jammu. I had come to the railway station to receive a retired regimental officer visiting the battalion. The train was late and I waited outside in my vehicle. Soon, a military police (CMP) vehicle arrived and a sergeant came out of the vehicle. Seeing me he said, 'Ram Ram Saab'. I acknowledged the greetings. He had been posted with me in the same station on three occasions over the last twenty years.

As we got talking, he said, *"Saab apko maine teen baar nazdik se dekha hai. Pahli baar 1983 mein jab aap Kota mein Aid-de- Camp (ADC) the. Dusri baar 1995 mein jab aap 5 GUARDS ke 2IC the, aur abhee Jammu mein aap 5 GUARDS ke CO saab hain. Yaad rakhna saab, ek na ek din aap jarur OC Provost (Military Police) ban ke aoge".* (Saab, I have served with you three times in my service. First time I saw you in Kota in 1983 when you were ADC to the divisional commander. The second time I saw you when you were the second- in- command of 5 GUARDS in 1995-96 and the third time in Jammu now when you are the CO. Mark my word, remember saab, one day you will definitely come as OC Provost).

I thanked him for his good wishes and laughed no end within myself. Well, as is the case, an officer is normally posted as OC Provost when he is either over the hill or his record of service does not promise any further promotion or on the last leg before retiring. Well, as far as the CMP NCO is concerned, from his point of view, his OC is the most powerful man in uniform. He can stop any vehicle and check anyone and file an offence report. His feelings were genuine and well meaning.

Well, my course mate laughed with me and the board result was forgotten for good. There are better things in life than to worry too much about smaller things like promotion.

Force Multipliers Par Excellence

The formation headquarters get their man power from earmarked units of a regiment in what is called the Extra Regimental Employment (ERE). The most important category of soldiers are the tradesmen that is - cooks, washerman, safaiwalas (cleaners) and clerks etc. It is generally believed that the units send the so called 'undesirable elements' on ERE or those with some problems which can be attended to on such postings closer home. You can make your own judgment.

One brigade headquarter at Dehradun received the following tradesmen on posting.

Office Clerk - Eyes allergic to artificial light (bulbs and tube lights).

Mess Cook - Low Medical Category, allergic to cooking gas.

Safaiwala (Cleaner) - Low Medical Category, allergic to dust.

Washerman- Body weight 45 kilograms, unable to lift the heavy military issue Coal Iron that weighs one hell of a lot.

Masalchi (Officers' Mess utensils cleaner) - Allergic to detergents.

When I was narrated this incident at Dehradun by the commander, I wished him good luck. Surely with this collection of gems, he could hardly be expected to launch into the battle. To add insult to injury, the ERE troops were from the commander's own regiment and he could not even complain or blame anyone.

Camp Commandant ki Jai

- *Camp Commandant: An appointment in a sector/brigade headquarters held by a junior officer. He is responsible for the functioning of the headquarter camp.*

It was the farewell visit of General Sekhon to a sector headquarters in the valley. Hectic activity was going on in the sector headquarters on the morning of the visit. Colonel Nijjar, the officiating commander wanted to give him a befitting farewell and was checking the finer details at the last minute. The General Officer Commanding (GOC) of the division had already reached and the Corps commander was expected shortly.

The senior Junior Commissioned Officer - Subedar Thapa, a diehard soldier and a simple soul comes and asks officiating commander. *"Saab chai ke baad teen bar 'Gen Sheikh' saab ki jai pukarna hai?"* (Saab, do we have to shout farewell cry for General Sheikh?) GOC who was within hearing distance laughed and corrected him saying "Saab, *Sheikh nahin 'Sekhon saab', Corps commander saab ka dharma mat badli karo."* (Not

Sheikh, General Sekhon. Don't change his religion). Thapa immediately noted down 'GENERAL SEKHON' on his palm and was even seen quietly practicing shouting *"Corps commander saab ki jai."*

As the Corps commander arrived, everything went as per plan. He addressed the men and had tea with them. As the moment of his departure came near, Colonel Nijjar looked at Subedar Thapa, who looked all eager and pumped up to perform the all important duty and was waiting for his signal. As the commander signalled him to go ahead, Subedar Thapa shouted at the top of his voice *"Camp Commandant Saab ki ...... "*. Everyone realized the goof up and it was not followed by *"jai"*. In response, there was only laughter from all sides including General Sekhon. Poor Thapa was red in his face hoping for the earth to give way. Later he was mortally scared to face the officiating commander. Nijjar having been with the Gurkhas for 27 years just laughed it off as one more incident of 'Humour in Uniform' involving a Gurkha.

Casualty Evacuation in Mountains, Avoiding Pain in the ...

During the field exercises in the army, there is a system of appointing umpires to conduct and monitor the progress of the exercise and bring out lessons at the end.

Recently commissioned, I was attending my first field exercise with my battalion as part of the brigade exercise. The exercise was set in the difficult high mountains of Sikkim. The umpire on the battalion was Major Dogra from the neighbouring unit. Well over six feet tall, he was an over blown heavy man with a treble chin, hanging cheeks, a big bulging paunch that concealed his belt all around his waist and massive thighs that rubbed against each other. Weighing over one hundred and thirty kilos and clearly unfit, he was struggling hard to keep pace with the exercise. Panting and sweating like mad, he was clearly miserable. Major Negi of my battalion looked at Major Dogra, completely haggard and done, sleeping against a tree some distance away,.

Evacuating a casualty in high mountains on a stretcher is a real pain. It takes four men to carry a stretcher up or down a treacherous track with no space to walk. It takes two sets of stretcher bearers to evacuate one man. By the time the casualty is evacuated, the stretcher bearers are half dead themselves. Those who have done it will know what I am talking about.

He said, "Evacuating such a heavy and unwieldy man on a stretcher in the mountains will be a real nightmare. The stretcher bearers are more likely to become casualty themselves." He then suggested, "When you get stuck with a man like him in the mountains, the medical support plan in the attack orders should read like this -

Hamle ke dauran agar koi jawan casualty ho jata hai toh use sikhe hue tarike se stretcher dwara regimental aid post tak evacuate kia jaega. Lekin, yadi Major Dogra casualty ho jata hai, toh samay na kharab karte hue, unhe blind samajh kar 'in situ' hee barbad kar dia jaega."

(During the attack in the mountains, if you have a casualty, he will be evacuated to the regimental aid post by stretcher bearers as per the laid down drills. However, should Major Dogra become a casualty, he will be considered a blind and destroyed in situ without wasting any time).

Digar Takes Them by Surprise

Major Gurinder Singh of Grenadiers had a batman, Grenadier Digar Singh Jakhar who was completely illiterate. Digar, conspicuous by his broken ears, was an outstanding wrestler. But that was the only field in which he could stake a claim to fame, and that was all he had done in life. The whole company knew that he was completely illiterate and could not even write his name; 'Angutha Chaap', as they are called. He would put his thumb impression for signatures for everything.

The troops were authorized rum as part of rations, and was called free issue rum. Rum was issued on fixed days, a much awaited occasion and normally coincided with meat issue day.

Normally a company rum issue stand consisted of three persons on a bench. A rusty patromax with a broken mantle provided semblance of light. It had to be delicately pumped every five minutes to keep going. The first in the team was a NCO with a rum issue register who recorded the details of individuals taking rum. The individual would sign against his name as a confirmation of having received his one peg. The second was the Quarter Master, who stood with a rum bottle and a peg measure for issuing rum. Some of them mastered the art of creating fourteen large pegs out of a bottle which has only twelve pegs, thus creating consumer surplus. A well rehearsed jerk of the hand while pouring rum into soldier's glass is all that it takes. The poorly lit corner for rum issue only helped matters. The third stands with a steel drum pouring water in the glasses. Salted dal fried from company cook house completes the menu. These days soldiers' bar concept has emerged and the troops have their rum in better environs.

Occasionally, the heavy drinkers would manipulate the non drinkers or absentees to make a few extra pegs for themselves.

The troops line up with their glasses and have their rum, normally in one shot.

Major Gurinder one day decided to check the rum issue record. He asked the company quartermaster whether his batman had his rum. Digar Singh rarely drank. The NCO replied that yes he had. The NCO showed him the rum issue register where Digar Singh had put his thumb impression for receiving his quota of rum. The officer told the NCO that he was lying. The NCO adamantly pointed at the thumb impression as the proof. The officer called for his batman and told him, "Digar show the quarter master how you would sign for your rum when you have one." His batman confidently took the pen and signed in barely readable English DSJ underlined with two dots.

The NCO stood dumb founded at the embarrassment. While the whole company knew Digar Singh as an 'Angutha Chhap' illiterate who could only put his thumb impression, unknown to the company, Digar Singh had learnt the art of signing in English without knowing ABC. The CQMH was shell shocked and caught red handed. The news of Digar's new found skills spread like wild fire and his stature in the company grew. No one whacked his rum again.

A Gentleman Loses a Gun and a Gentleman Steals it

- *Junior Commissioned Officer (JCO)- A middle rung rank below officers and above NCOs. Equivalent of a Warrant Officer or Master Sergeant in foreign armies.*

Sometime in 2007, a Sikh JCO was carrying two carbines in his hand bag, travelling to another city. Waiting for the train at the railway platform, he acquainted an old

Sikh passenger. He requested him to look after his bag while he went looking for his reservation from the reservation chart.

When he returned, the person had vanished with the bag with carbines. It was a major act of neglect and violation of orders in handling of weapons.

One staff officer at the corps headquarters, reporting the incident to the Brigadier General Staff said, "As per initial report, the carbines have been stolen by a Sikh Gentleman." And who lost it? He was asked, "Another Sikh Gentleman," he replied seriously.

I kept wondering what could be so gentlemanly about two surds; one stealing weapons, and the other losing them.

Smart Parry

Parry was a 'Direct Entry' course mate, one of the seven commissioned in Brigade of the Guards in Jun 79. We landed at the Guards Regimental Centre for a fifteen days attachment after commission.

Along with speed marches and regimental history, weapon training formed an important part of the curriculum. The weapon instructor was a clued up fellow whom we called ustad.

Parry seemingly had whiled away his time as a cadet at the Indian Military Academy and knew very little about weapon handling. The ustad took no time in realizing that friend Parry was all at sea.

We used to sit on the squad post in order of battalion seniority. Parry was second last i.e. second from right. The ustad would call Parry very often to give weapon handling demonstration and Parry would cut a sorry figure every time.

After two days, Parry, clearly fed up, says, "This ustad thinks he is too smart, just watch what I do tomorrow."

We forgot about this as one more of Parry's casual meaningless comment. The next day we all sat on the squad post for the weapon class. Sure enough, when it was time to call an officer for a demonstration, the ustad said, "Second from right." Parry confidently got up, raised his right hand and showed index finger, "Ustad, I am hurt." (Index finger is used to do bulk of weapon handling activity- cocking of the weapon, trigger operation, magazine filling etc).

Parry's right hand index finger was heavily bandaged & half red. Since we used to spend most of the day together, we wondered when and how did he injure himself. None of us had a clue.

Ustad said, "OK sit down". He asked someone else to give the demonstration.

Over the next period, it happened a few more times. Whenever the ustad asked 'Second from right' (Parry) to give the demo, Parry quietly raised his injured finger and the ustad would ask him to sit down.

A short tea break followed and Parry said, "See, didn't I tell you. I will fix this damn ustad. He thinks he is too smart- after me all the time." No one asked him how he had got hurt.

We settled down on the squad post for the class again. This time the ustad asked,

"Class sochen aur bataen LMG ki magazine mein kitne round ate hain?"

(How many rounds are filled in a LMG magazine?)

And he said, "Second from right", which was Parry.

Parry confidently raised his bandaged finger, *"Ustad, ungli mein chote."* (I am hurt).

Well, well we had a great laugh as Parry stood expressionless, beaten by the ustad at his own game. As for his injury, there was none. It was bandage and red ink.

Toeing the Line

I was Aid -de-Camp to a divisional commander (GOC) in early eighties. Suffering from dental ailment, the GOC visited the Military Dental Centre. The divisional medical advisor (ADMS) and self accompanied him. By the time he had finished, it was past office time and we headed back to drop him at the Flag Staff house.

As the staff car halted in front of the house, the GOC's dog, Gompu, a small black Tibetan breed with long hairs came out running to greet his master, wildly wagging his tail. It was a daily affair. The GOC would pat Gompu and say something. Satisfied, the dog would amble back into the house behind the GOC.

It was a hot and sunny day and as usual Gompu came out running to receive his master. The GOC said in Hindi, roughly translated would read, "Come on Gompu, it is bloody hot go inside." The ADMS then said this, translated to English would read "Come on Gompu, it is too hot outside, go inside or you will get dark." Gompu was jet black. With age he was beginning to go grey on the head. He would have been very happy to turn black if he could help it. So much for toeing the line.

Never Leave Your Wife in the Lurch

12 GUARDS had recently won a major formation championship. The battalion decided to beat it up in the mess, as it happens on winning a major competition. They also invited the brigade commander and the staff to join the party.

On the way back from officers' mess, the brigade commander remarked, "12 GUARDS is a bloody good battalion." The Brigade Major butted in "Sir, this competition is fine, let us see how good they are in battle preparedness as well." "Fair enough," said the commander, so it was decided to put the battalion through a mob scheme rehearsal the same night without notice.

The beat up finished well after mid night. The officiating commanding officer (CO) had been recently married. His wife coming from a civil background did not understand the nuances of the military. The CO and his pretty wife had just about gone to blissful sleep when his rickety field telephone came to life. He wondered what was so important at this beastly hour. The duty clerk read a log message from the brigade headquarter.

From: HQ 280 Armoured Brigade
To: 12 GUARDS

Personal for commanding officer (.) mob scheme practice (.) battalion to mob for operations immediately with hundred percent equipment and manpower (.) unit to report when ready (.) commander will inspect the unit (.) ack

On mobilization, called mob scheme, the units are required to move out combat ready in full battle gear in a very short period. This is not feasible unless the unit drills are well taped up. When done without prior notice, it is a litmus test. Time is of essence, hence the hurry. It is seen as a measure of unit's battle preparedness and thus taken seriously by all.

The CO who was staying in the officers' mess threw off the bed sheet which uncovered his wife. He jumped off the bed, and bolted off in vest and pajamas to inform the officers to get going. Seeing her husband, the newly wedded wife was in near panic. She was out of the bed like uncorked champagne and followed him asking what had happened, but he was too occupied to reply. The CO informed all officers, rushed back to the room and was quickly changing in to battle gear throwing his night suit and pajama all over the room.

His wife, in near panic, kept asking him what was going on. But he was hardly listening, busy planning his next move. When he was about to leave, his very worried wife asked again,

"What is the matter, why are you rushing around? Why are you wearing combat dress at this late hour?" The husband replied, "You stay here, I am going, we have a mob scheme on."

The lady by now near hysterical, said, *"Arre yahan mob aah raha hai aur aap mujhe akele chor kar ja rahe ho."* (My God, look at you. A mob is coming here and you are leaving me alone). The CO could not keep from laughing his guts out and made her feel more miserable.

Empty Vessels Make More Noise

A commanding officer telling a Second Lieutenant who was talking too much without making much sense. "If ever there was a trophy for talking maximum amount of nonsense in shortest amount of time, you would have got it without contest."

Improvisation thy Name is Indian Army

• *Bangalore torpedo. A bamboo shaped explosive used to blow up wire obstacles or bunkers. It is about six feet long and three inches in diameter.*

Indian infantry soldier is a wonderful guy. He always delivers in crisis. Coming essentially (not as much lately) from rural areas and moderately educated, he gets the job done once he knows what needs to be done. Sometimes improvisation can be taken to ridiculous limits either by design or default.

Serving in J&K in the eighties, the Magazine NCO (responsible for ammunition) of our neighbouring battalion reported to the Quarter Master (QM) that one service Bangalore torpedo was not traceable in the magazine.

The QM, a Gurkha officer jumped out of his seat. "What nonsense?" he shouted. "How can a Bangalore torpedo be missing. I want the whole battalion turned upside down and in next three hours I want that traced."

It was a bloody serious matter. Loss of such a major item of ammunition can neither be accepted nor covered, worse, it is an embarrassment beyond description. A massive search was launched in every nook and corner without success. It was decided to wait for sometime before reporting the matter as it would bring a bad name to the unit. One company of this battalion was on training about eighty kilometres away at Sonamarg. It was the only company which was not turned upside down.

The adjutant of the battalion visited the company to see how the training was going on. The jawans of the company invited the officer for lunch in their tent. The officer went in and was made comfortable on a ghee tin with a blanket. Moving into a dark tent from a bright sun, his eyes slowly adjusted to the darkness and he began to see what all lay inside. And then he noticed that the centre pole of the tent was smaller and thinner than the others. It was rested on an inverted bucket to increase its height. As his eyes adapted to the darkness, he was shocked to see that the centre pole was none other than the missing Bangalore torpedo. Right next to that was the lunch being warmed on a stove burning in full glory .The Bangalore torpedo had been missing for weeks. Many a lunch and dinners had been cooked next to that explosive that could blow up a concrete bunker. Just proves, yes there is God somewhere.

The battalion breathed a sigh of relief. An embarrassment was avoided and a disaster averted, thanks to the God almighty.

Attack across a canal: Improvisation, thy name is infantry

Expecting too Much?

The military exercises for the troops tend to become meaningless if they are not involved and kept informed. They invariably become an exercise for officers with men just going through the motions without deriving training value worth the name.

In a division, the General Officer Commanding (GOC) was particular that the troops know what is happening all the time so that maximum benefit is derived from the exercise. Normally during an exercise if you asked a soldier what is happening, he would reply *"Saab scheme chal rahi hai."* (Scheme is on. Scheme is a slang amongst troops for exercise with troops).

The GOC passed specific instructions that during the coming exercise he would not like to hear this term 'scheme' from any soldier. He would speak to the troops to check that they know what exactly is going on and what their role is. Every unit made sure that the troops were briefed and kept informed all the time about the progress of exercise as to what has happened, what is happening and what will happen. The GOC who was moving around spoke to a Junior Commissioned Officer (JCO) and asked him,

"Saab, aap batao kya ho raha hai." (Tell me what is happening).

The JCO replied *"Saab pahle paltan ko attack ka Warning Order mila. Phir Brigade commander saab ne apne 'O' Group ko hukam diye. Phir CO saab ne warning order paltan ko pass kiya. Paltan mein battle procedure chalu hua. Paltan ke CO saab ne battalion O group ko attack ke hukam diye aur paltan ne dushman ke defences ko contact karne ke lie advance jari kiya. Battalion ne assembly area mein aa-kar quick defence liya.* (First the brigade commander issued his orders, after which, the CO issued his warning order. The battle procedure started in the battalion. The orders for attack were issued and the battalion started the advance to contact. The battalion then took up hasty defence in the Assembly area).

"Bahut achha saab phir kya hua?" (Very good and then what happened?) asked the GOC.

"Saab, phir pakke hukam hasil hue, aur paltan ne jordar hamla karte hue objective par kabja kar liya." (Final orders were issued and the battalion captured the objective after a ferocious attack).

"Bahut achha saab, phir uske baad kya hua?" (Very good and then what happened?) enquired the GOC.

"Saab phir paltan ne re-org ki karwai ki aur defence akhtiyar kiya,"

(Battalion occupied defences after re-organization drills), replied the JCO confidently.

"Bahut achha saab. Ham khush hain ki aap ko pata hai ki

kya ho raha hai. Ab aap batayen ki age kya hone wala hai," (I am happy that you know what has happened, and now tell me what will happen next), said the GOC.

"Bus saab, ab is ke age 'scheme' chalne wala hai," (That is it, after this will be the scheme), replied the JCO.

The GOC looked at the brigade commander, the commander was looking at the commanding officer who was already looking the other away. I guess the GOC was expecting too much.

A Day the Commanding Officer Won't Forget in a Hurry

4 Guards (1 Rajput), a battalion with great history was out on exercise. As the culture was those days, the battalion was walking back to its location after the exercise. A small route march of about two hundred kilometres covered over a few days. Everyone including the commanding officer (CO) was on 'Number 11 bus', a term we use in the army for foot slogging. After the day's trudge, the battalion would bivouac somewhere on the barren road side. Everyone slept in the open. The only privileged man for whom a small forty pounder tent was pitched was the *'Gaon Burha'* as the CO is known in the battalion. The administration vehicles would come up ahead and cook dinner for all ranks. A rickety vehicle called the F- Echelon 1 Ton carried the CO's tent and a folding field cot- issued on commissioning, but rarely, if ever used, for it was specially designed to give you a life time backache. Amongst other utilities in the vehicle was a special piece of military hardware called the *'Thunder Box'* (now extinct). It was really a field WC of the modern day, called commode in the military parlance. It was designed like a foldable wooden chair

with a removable metal pan placed in it. From a distance it looked like a chair.

The unit Subedar Major with imposing Rajput moustaches was an extremely sincere man. After the day's march, he would take stock of the situation, check out the battalion, come to the CO and give him a detailed report. More detailed than the CO desired. Nevertheless, he would listen patiently as the Subedar Major went over his notes, frequently looking up.

The CO having gone around himself would come to his tent pitched some distance away. There, his batman, now called buddy, laid out the clumsy canvas field chair outside the tent- another back breaking hardware specially fabricated for the army. The CO would sit on the chair in shorts, reading a newspaper, as his loyal buddy gave him a hot mug of not so hot tea. Yes, the military mugs were designed that way. The high technology metallurgy made the mug too hot to put to your mouth while tea in it became cold in no time.

One day after the route march the Subedar Major came to the CO for the evening report. He saw the CO sitting with a newspaper behind the F Echelon 1 Ton. A little unusual, the Subedar Major thought. The CO was taken aback at seeing him. The Subedar Major followed his daily ritual with a 'Ram Ram saab' accompanied by a clumsy salute and started his long discourse of the day.

"Saab, Paltan ke hathiyar, jawan sab theek thaak pahunch gaye hai. Teen jawan ke shoe bite ho gayya hai. Unhen MI Room mein bhej diya gaya hai. Alfa company ke senior JCO ko bukhar hai unka bhee treatment chal raha hai." (Saab, all well in the battalion. All officers, men and weapons are fine. Three men have suffered shoe bite and the Senior JCO of Alfa company is suffering from fever. They are all being treated in MI Room).

"Theek hai saab, baki report baad mein dena," (OK saab, very good. You can give me balance of the report later), said the uneasy CO politely.

Unmindful of the CO's suggestion, the Subedar Major pressed on with the report, intermittently looking straight in the CO's eyes.

"Aur saab, sab company ka langar set ho gaya hai aur khana taiyar ho raha hai. Officers saab log jawanon ke administration dekh rahen hain. Aur------" (Saab, all companies' cook houses have been established and the food is being cooked. The officers are in the companies looking after the men's administration. And -------)

"Theek hai- theek hai saab, baki report baad mein dena. Mein apko bula loonga," (OK, OK that will do. Give me the balance of the report later. I will call you), interrupted the CO impatiently.

Not to be deterred, the Subedar Major continued with all his characteristic seriousness. *"Ek 3 Ton ka silencer toot gaya hai saab, bahut halla kar raha hai. URO ka vehicle mechanic usko dekh raha hai. Aur-----* (The silencer of a 3 Ton vehicle has broken down. It is creating one hell'va racket. It is being repaired by the workshop mechanic. And--------)

"Theek hai- theek hai saab, abhee mein busy hoon baki report baad mein telephone par dena, mein app ko telephone kar doonga. Mein apko bula loonga," (OK, OK that will do. Right now I am busy. I will give you a call later and then you can tell me what is left), interrupted the CO definitely annoyed by now. The CO unsuccessfully tried to end the report many times that would have taken almost fifteen minutes. The Subedar Major, used to giving long detailed reports carried on, at his leisurely pace, till he had finished- to his satisfaction. The CO's interjections to wean him off failed miserably.

"Mere liye koi hukam saab?"(Any orders for me sir?) asked the Subedar Major.

"Nahi –nahi saab, koi hukam nahi , abhi aap jao, hoga toh apko telephone kar doonga." (No, none at all. You can leave. If there is anything, I will give you a call), replied the CO

impatiently holding the newspaper firmly.

"Bahut achha saab, toh kal subah paltan subah paanch baje march karne ke liye taiyaar rahegi aur apko report lag jaegi saab, Ram Ram saab," (Very well saab. The battalion will be ready to march sharp at five in the morning and I will give you the report. Good day saab), said the Subedar Major.

The CO was certainly relieved as he saw him walk away. In all, the Subedar Major had taken almost fifteen minutes to complete his report. As he was walking back he met the CO's batman.

"Arre chore, tune aaj CO saab ka chair bahut door, gaari ke peeche lagaya, kyon?" (Soldier, today you placed the CO's chair far away, why?") asked the Subedar Major.

As the Subedar Major heard the batman's reply, his eyes began to pop out and he fidgeted uncomfortably.

The Subedar Major's family jewels must have crashed with a bang. All the while when he was giving his marathon report, the CO was his captive audience sitting on the 'thunder box' with a bad stomach! Tally ho.

The CO's impatience made sense to him now. I am told he familiarized himself with all models of commodes in use in the battalion after this episode. Can't really blame the Subedar Major. I guess the CO had set a standard routine and he got ambushed- he was reading a newspaper at the wrong time and the wrong place. The newspaper may have been part of the problem to begin with. In the end, it came in handy. Without it the old man's secrets would have been laid bare, nice and proper.

Computer and The Commander

Computers were introduced in the army at unit levels around 1999-2000. By then my generation, and earlier ones, had reached a stage where someone else did the leg work for you. As a result most of us remained computer illiterate beyond switching the computer on and off.

We were part of a control headquarter during the divisional war game. The team was headed by, let us call him Brigadier Choudhary, the artillery brigade commander, who was the Chief Controller. Over two days, and after a lot of effort we prepared a power point presentation for briefing the General Officer Commanding (GOC) of the division on the conduct of war game. Brig Choudhary, a wonderful man, asked us to come early the next day so that we would be ready for the GOC in time. My other colleague Colonel Chatterji, of the artillery brigade, was a computer wizard who had done all the leg work on the presentation.

When we opened the presentation, Chatts shouted, "Oh f***s man, the presentation has gone. Someone has over written an old file on our file. What do we do now?".

This meant that a very old draft of the presentation was back. All our days' efforts had disappeared with a click of a mouse button, and with it all the amendments the commander had painstakingly incorporated.

I said, "There is no time to do anything now. Leave it as it is, no one will notice the faux pas." Unnoticed, Brig Choudhary had come in and stood behind us as we stared at the computer.

"What happened?" he asked.

"Nothing serious," I said, wanting to avoid unnecessary panic.

"No, I heard Chatterji say there is some problem," he said.

"Sir, our final presentation has gone. Someone has overwritten an old file on our presentation file," said Chatterji.

Choudhary was confused and devastated. He then said, "What are you chaps doing yaar. Overwritten? Let me see."

He then craned his neck over our heads to see the screen. Apparently relieved, he said, "I cannot see any over writing there."

We were equally relieved and the briefing went off well.

IMPROVING TECHNOLOGY THRESHOLD IN ARMY

Texan and the Tally Ho

Major Shyam Kumar of 5 GUARDS, later a corps commander, was a flamboyant upcoming officer who often used jargons to convey his point. He was a Brigade Major (BM) of a brigade that was undergoing an exercise on attack in mountains in the East. He was talking on radio with a battalion commander from Haryana who was all set to launch his battalion into attack. The commanding officer (CO), a simple soul, did not understand the nuances of English too well. The radio conversation proceeded like this:

CO: "Tiger 14 for Lion, message, Over."

Kumar: "Lion listening go ahead, Over."

CO: "Tiger for Lion, we have married up with Serial 2 at RV and proceeding to the 'Forming Up Place'. Shall launch attack as planned. Any further instructions, Over."

Kumar: "Lion for Tiger, so far so good, proceed on your mission as per plans and report completion. Tally Ho, Over."

CO: "Tiger for Lion, rogger, Wait."

Kumar waited for the CO on the set. After some time the CO was back on the radio.

CO: "Tiger for Lion, reference your last message, code word 'Tally Ho' not understood, clarify, Over." (Tally Ho was Kumar's style of saying "All the Best").

Moon-a-Boon

The deserts due to the absence of landmarks can be a navigator's nightmare. During an exercise with troops one company got lost and could not marry up with the battalion for the attack. To make matters worse, the company commander was also out of radio communication. Finally, when the radio contact was established, the commanding officer came on the air to check-out the company location.

"Kilo 15 for Kilo 4 report your location, Over."

"Kilo 4 for Kilo 15, my location, reference moon six 'o' clock. Request your location, Over."

"Kilo 15 for Kilo 4 same, reference moon six 'o' clock, Over."

"Kilo 4 for Kilo 15 then we are close by, wave your hand, and should be reaching soon, Out."

Need not be mentioned that they never met for next eighteen hours.

The Notional Notion

Indian soldier is one of the finest soldiers on earth led by the finest young officers. Well led, he will follow the officer unquestioningly to his death. He can be lively and extremely humorous under trying conditions.

During the winters of 1985 one division was undergoing a collective exercise with troops in the deserts of Rajasthan. To preserve the expensive equipment from undue wear and tear, equipment participation was scaled down. A small skeleton force represented a larger sub unit, one tank or a one-1 Ton wheeled truck with a log positioned on the bonnet represented a troop of three tanks. Three such vehicles represented a tank squadron. Similarly, one artillery gun would represent six guns and so on. In that exercise, even infantry, which normally does not have such a privilege, represented a non-existent force termed 'notional'.

A Gurkha battalion was tasked to capture a feature. The battalion prepared well for the attack. Proper orders were issued and thorough rehearsals were carried out. Quietly the battalion occupied the Forming Up Place (FUP) and closed in with the objective. Within two hundred yards of the objective, the fired up Gurkhas shouted their battle cry that rented the air. Khukri in one hand and the rifle in the other, the Gurkhas charged the objective only to find that the feature was not occupied. Gurkhas got down to digging to occupy defences on the captured objective.

In the distance a cloud of dust approached the Gurkhas. A sentry drew his khukri and was prepared for the intruder. As the vehicle closed in, a white flag of the exercise umpire became visible. The umpire drove up the sand dune and declared the attack to be a failure. The Gurkhas argued that there was no one defending the feature when the attack was launched, how can the attack be a failure? They were told, "No, there was a battalion here defending the feature, except that it was notional. A battalion attack on a battalion position cannot succeed, period." The commanding officer had a tough time explaining to his men as to why the attack was a failure. The Gurkhas reluctantly filled their trenches and trudged back to where they had started from. The word notional caught up with the troops quickly and became an object of fun and laughter.

One young Engineer officer who was part of the umpire team that had judged the Gurkha attack a failure was an umpire on the same battalion. He noticed Gurkha johnnies sitting under the bushes and making merry, apparently eating lunch from their mess tins. On getting closer he saw that they had empty mess tins in their hands and were acting as if they were eating food. He approached them and asked:

"Kancha kya kar rahe ho?" (What are you doing?)

"Huzur khana kha rahen hai." (Sir we are eating our lunch).

"Lekin aap ke mess tin toh khali hai, usme kuch bhi nahin hai." (But your mess tins are empty, there is no food in them).

"Nahin saab hai, lekin notional hai." (No it is there, but it is notional). The officer had a good laugh.

"Achha, lekin kha kya rahe ho, kancha?" (OK, but what are you eating?) asked the officer.

"Saab dal bhaat kha rahe hain," (We are eating dal and rice, a favourite Gurkha meal). replied the Gurkha soldier.

The officer then said, *"Arre kanchon, dimag lagao. Agar notional khana hee khana hai toh chicken, meat aur machhli khao, dal bhaat kyon khaa rahe ho."*

(If you have to eat notional food, why eat rice and dal, why not eat chicken, mutton and fish).

The Gurkhas liked the concept and then the company CHM blew a whistle and announced loudly,

"Aaj mere taraf se shaam ko bara khana hoga. Shaam ke khana ka menu mein meat pulao, chicken masala aur rum ke saath macchli hoga. Aap sab shamil honge. Yaad rakho jo bara

khana mein de raha hoon, woh notional hoga. Sara kharcha mein uthaonga."

(Today I will host the dinner. The menu for the dinner will be chicken, meat and fish. You are all invited. The dinner will be notional and I will bear all the expenses).

Caught you Hearing

Some officers on becoming permanent low medical categories lose interest in service, especially when the category renders the future promotion prospects dark. With no prospect for promotion, some tend to become what in the army is termed 'FF' or freedom fighters. The term signifies being the son-in-law of the army and living off the land.

One officer who was a Low Medical Category (LMC) for a hearing problem had been avoiding military duties on this pretext. One day during an officers' conference he decided to project his case to the commanding officer (CO).

"I cannot attend the morning PT parade because when the second-in-command gives the order for the battalion to come to savdhan (attention) I cannot hear. I continue to remain in veeshram (at ease) when the whole battalion is in savdhan and the troops laugh at me.

I cannot attend the commanding officer's conference because I cannot hear properly and it becomes embarrassing. I cannot travel in a 1 Ton vehicle (a noisy uncomfortable military vehicle, no more in service) because it makes a lot of noise and it will damage my ears further. I therefore need a jeep (a rare commodity in an infantry battalion, and a rare privilege those days) to move around. The CO heard him out and then said very softly, "One officer is required to go on temporary duty to Delhi from our unit, any volunteers." It was said so softly that half the officers sitting in the first row did not hear. He, sitting in the third, without wasting a second, said, "Yes sir, me."

The CO smiled, looking towards him from the corner of his eye as the officers burst out laughing. The officer never again complained of his impaired hearing.

The Jats and their Kheer

The Jat troops are known for their typical traits. Their accent, referred to in the army as the Texan accent is misunderstood as rustic, rude and blunt. Generally vegetarian, they are fond of the traditional Indian pudding made of rice and milk, called Kheer. Their collective weakness for tel-maalish (oil massage) and desi ghee is well known in the army. Every Jat soldier returns from leave with a huge tin of desi ghee.

A Jat battalion had won the brigade Mortar shooting championship. The Mortar platoon had celebrated their victory with Kheer. A lot of it was left over and no suitable place to keep it. The brigade commander could not be present for the competition due to some commitments. While the platoon was all set to fall back to unit's permanent location the next day, the battalion got the message that the commander would like to see their firing. A big problem indeed thought the whole platoon.

The platoon commander, a young Captain, came over to the platoon early the next day and gave them the news. The soldiers, normally bubbling to fire for a VIP looked at each other saying nothing. The Captain seeing the lack of enthusiasm asked them, "Are you boys ready to fire?" trying to pep them up.

The Mortar Platoon NCO said, *"Ham bilkul taiyyar hain*

saab, chaar Mortar se cover kar lenge saab." (We are ready sir. We will manage with four Mortar tubes).

"Kyon, poore chhe Mortar kyon nahin?" (Why not all six Mortar tubes?) asked the Captain.

"Baaki dow mein toh Kheer bhari hai saab." (The other two Mortar tubes are full of pudding sir).

Doctor's Plight in Family Planning

I don't know if it is true, but it is a very popular joke about the Gurkha Regiment doing rounds in the Indian army. During the mid eighties there was a big drive on family planning. The condoms had just entered the scene. Troops were educated on benefits of small families and in variety of family planning measures. The Regimental Medical Officer (RMO) explained the use of condom by pulling it over his thumb. *"Any doubts?"* asked the doctor. None had any doubts on its use after a very practical lecture and demonstration.

Those days the Gurkha troops from Nepal invariably used to avail their leave once in two years because of long travelling time. One soldier returned from four months leave and reported to the doctor. What's the problem asked the doctor?

The soldier innocently explained; he already had three children and did not want anymore. So he had used condom during his leave, still his wife conceived and got pregnant.

"Couldn't be," said the RMO. "How did you use the condom?" he asked.

"The way you taught us in the Sainik Sammelan saab," and then slipping a condom over his thumb, "Exactly like this," said the soldier. The RMO had taken too much for granted.

South Paw

I was posted in a division headquarters in the eighties. We were chatting in a group during the weekly central tea break. One simpleton happy go lucky Gurkha officer from the ranks, Capt Chhetri, was narrating his experience of a boxing bout as an IMA cadet. He had knocked out a big African cadet, who was a potential Olympic boxer. He said that his opponent had a clear advantage. Apart from being an outstanding boxer, he was tall with good reach and went on to describe the bout.

He stood with clinched fists and left shoulder leaning. The way he was standing and punching the air, it appeared to me he was a lefty.

Gesturing as a lefty, I asked him, "Are you a south paw?"

"No, in boxing I use both my hands and paws," he said, moving both his hands up and down viciously. (South paw is a left hand dominant boxer. Chhetri was an outstanding boxer who knew no theory of boxing).

The Long and Short of it

A unit has more than a dozen officers. Every officer is an individual with different traits and personality and contributes to the strength of the unit in his own way.

In the seventies, one battalion had an interesting officer, let us call him Major Mannu. He was very fond of laying bets on almost anything. Surprisingly he would win more often than he would lose. One area where he rarely lost a bet was the physical dimensions of any item. That had begun to make the officers interested. The bets would normally be related to the height, weight and chest of items of furniture or any other piece of equipment. Sitting in the officers' mess or anywhere else, he would casually slip in "What do you think will be the height of the table or the height of the bulb from the floor or something else like this. Every time he chose a different place, audience, time and the issue of bet. He had developed the knack of putting it across in a manner that someone or the other got roped in unknowingly. A discussion would then follow, invariably ending in a bet with many beer bottles at stake. The object would be measured and whenever the stakes were high he would win. If the stakes were inconsequential, he would lose, and that maintained his credibility. He had been winning the heavier bets far too often, beating the law of averages. It was too good to be true. The officers marvelled at his judgment of estimation but were suspicious, but could not nail him. Yet the delicate moment he chose, invariably after many drinks had done the rounds, caught them off guard.

One day the battalion was having a party in the officers' mess. When the party was at its peak and every one generally glad crowding the bar, Mannu quietly brought up the issue of the length of the leg of the bar stool. "Let us see if the Guardsmen can think straight after booze." Heavy bets were laid by all officers totalling more than forty beer bottles. Mannu was happy he had succeeded in encashing on the 'tipsy Guardsmen' in the middle of the night. Lot of 'halla- golla' and the ladies joined the fun. Mannu then said, "Let us measure the length of the leg of the bar stool."

Everyone guesstimated the length. Mannu skilfully guided the mob to estimate the length to be less than 35 inches. When bulk of the officers fell in the trap and said that the leg was less than 35 inches, Mannu said by his estimate it should be more than that. Of the eleven officers, eight participated in the bet, two abstained and the commanding officer (CO) was left out to be the adjudicator. He watched the fun in silence and amusement.

The CO was requested to announce the judgment. He

asked one of the abstainers to measure the length of leg. The officer rolled up his sleeves, jumped in the ring and with stretched palm measured the leg of the bar stool.

"More than 35 inches," he announced, "Major Mannu wins the bet hands down. Ladies and gentlemen, give him a big hand."

Major Mannu, who always looked confident throughout yelled in excitement asking the barman to serve a round of beer and send the rest to his home.

The game was not over yet. The losers contested the decision in unison.

"You can't measure the length this way," said one.

"Get a measuring tape," shouted a few others. A mess waiter was sent to find one, and surprisingly, he found one rather fast.

As everyone crowded around the bar stool, the abstainer measured the length. The Mess staff in full regalia surrounded the officers and joined the fun.

The officer put one end of the tape at the top of the bar stool and went downwards. "Leg length 34 inches ladies and gentlemen, the old judgment stands cancelled, I am afraid Major Mannu loses the bet and all the beer," declared the adjudicator.

The whole mess went up in cheers as the officers celebrated victory.

"Stop, shouted Major Mannu. Impossible, it can't be true. It must be measured again." It was measured again, 34 inches shouted the adjudicator. Another round of celebrations. The CO declared the result as final.

Major Mannu snatched the tape and measured it himself, alas, the leg refused to grow however often he measured. Suffering from shock and awe, he slumped on the same bar stool.

"Hey, let's finish the beer here," shouted one officer.

Everyone grabbed a bottle and drank with much laughter and celebration. Mannu was also offered one, which he took with total lack of enthusiasm.

The party broke in the wee hours of the morning. Major Mannu was too tipsy and shocked and awed to walk back home. He was driven home and dumped on his bed, his wife in attendance.

What had actually happened was that on the day when the party was scheduled, one subaltern saw Mannu measuring the length of the leg of bar stool. So that was the secret of his betting success, he thought. Clever man, he put two and two together and reported the incident to the second-in-command. A quick clandestine meeting of the officers was held. It was decided to cut the legs of all four bar stools by one inch.

Everyone was briefed on his part for the evening. Two officers were nominated to abstain from the bet, other seven were asked to act as if in high spirits. The CO was not informed. Everything fell in place and the rest is history.

Next day Mannu was reported sick in quarters. When he came to senses he could not figure out how he lost the bet that resulted in one big dip in the pocket. It was much later that he learnt of the trick.

When Maser Juxtaposed Laser

In a promotion examination for promotion from Captain to Major the students were required to write a short note on MASER. One student who had only heard of LASER got up and told the invigilator that there was a typing error in the paper and that it should be LASER and not MASER. The invigilator, equally brilliant, agreed and proceeded to the other hall and announced the correction. He then came back to his own hall and told the senior invigilator to announce the amendment. He was a sharp guy and told him what the hell he was up to, that there was something called MASER also and that the question was right. The mistake was corrected.

The student officer taking the exam who had started the confusion, now looked for an opportunity to get a hint from the officer sitting next. When he got one, his friend, equally brilliant, prompted "MASER is a type of an antibiotic." The officer wrote a long short note on antibiotics.

Ramban; Variety is the Spice of Life

A large part of the Indian army is deployed in the state of J&K for more than one reason. There is a regular movement of soldiers from Jammu to the valley through a series of transit camps starting from Jammu and vice versa. En-route there is a transit camp at Ramban. In the earlier days this transit camp was meant only for providing lunch to the transients moving up and down and did not cater for night halt for either officers or men.

The transit camp was famous for its excellent food. A convoy got stranded at Ramban due to landslide and was forced to halt there for the night.

An excellent multi course lunch was served. All officers

and families heartily appreciated the sumptuous lunch and complimented the OC Transit Camp, Mess Havaldar and the cook. Since there was no accommodation, officers and families were accommodated in the drawing room, dining hall, corridors and wherever the space existed.

Come night and the crowd gathered again for dinner. A sumptuous dinner followed, albeit exactly the same menu. Again rounds of compliments to the transit camp team. Some officers noticed the repetitive menu.

The road to Srinagar remained closed for the second day. The next day treated the officers to a steamy hot lunch, but exactly the same menu of the last two meals. Another round of compliments was passed around to the transit camp team. The officers however had begun to get fed up of the monotony of the menu.

Come the night and steamy hot sumptuous dinner followed, and yet again, exactly the same menu. Officers thought it was time to get into the act. One officer called the Mess Havaldar and said,

"Hamme yahan do din ho gaye hain. Dow bar lunch aur dow bar dinner kha chukey hain. Har bar ek hee menu. Dow din mein ek sabjee bhi nahin badli. Yahan menu badli nahi hota kya?"

(We have been here for two days. We have had lunch and dinner twice and the menu has been exactly the same in all four meals. Don't you change the menu here for God's sake?)

"Nahin Saab, yeh Ramban Transit Camp hai. Yahan upar niche ane jane walon ko sirf lunch diya jata hai.Yahan menu nahin, yahan afsar badli hote hain."

(No sir, this is Ramban Tansit Camp. This transit camp is only meant to provide lunch to the officers going up and down. Here, it is not the menu but the officers who change).

Traditions Die Hard

One Grenadiers battalion was located in a high altitude area. Major Sangwan, a company commander had been on the post for almost six months. During this period another unit had come in and was located right next to their base.

Sangwan came down to the base for a break. As per military customs unit quarter guards sound the bell for every hour to indicate the time of the day. At exactly eleven in the afternoon, the quarter guard sentry started sounding the bell. This officer sub-consciously counted the gongs. To his surprise, the gongs continued past eleven and he thought he had counted till twenty two or twenty three. He assumed the quarter guard NCO had probably lost count half way and started all over again.

The next day Sangwan was telling another officer, "After so many years of service, I learnt only yesterday that in quarter guard the NCO could start sounding the bell all over again if he lost count half way. Yesterday, my quarter guard NCO sounded the bell twenty two times at eleven in the morning."

He was really amused, and I guess a bit embarrassed, when someone told him that it was not the case of missing the count of the gongs. The truth was that there was another unit's quarter guard right next to their own, which always sounded its gongs after the first had finished theirs, and that is how he heard twenty two of them.

Well, maybe they could have done with just one quarter guard sounding the bell, but traditions die hard, never mind if it leads to confusion.

Tane Toh Lath Gaar Diya Saab

Major Sangwan who has been mentioned in the episode above proceeded on leave. Thanks to places where they are normally deployed, officers mostly land up travelling without rail reservation. In the train he could somehow manage the top berth by a joint operation with his sahayak, also proceeding on leave with him. When he got up in the morning the train was at his home city Meerut. On enquiring, he was told that the train had been there for a long time and could be leaving any time. He hurriedly packed his bedding and stepped down the berth only to find that his shoes were gone. Unfortunately he was not carrying a spare pair. A crisis was at hand. An officer of the Indian Army could not be walking barefoot on the station. Something had to be done. He then saw his sahayak coming back to the compartment. Soon the jawan realized what had happened and his military training told him what needed to be done. He told his officer to wait and he would be back soon. The officer waited and as the train started to roll out, the officer saw his sahayak.

"Kyon chore sab theek hai na?" (Hope all is well?)

"Aa jao saab, sab theek hai." (Yes, come out all is well).

The officer confidently stepped out bare footed from the moving train. The jawan placed a fine pair of shoes in front of him, a perfect fit. The officer wore the shoes and walked off, followed by his faithful batman achieving total surprise. Later, the soldier told his officer *"Saab taane toh lath gaar diya,"* (Sir, you have done a great job). Well, it was the soldier who had done the smart trick. Just too bad, someone else would have to manage another pair of shoes somewhere else, some other time.

Sahayak; That Wonderful Institution

Batman or sahayak in the army is an institution by himself. He has undergone many a name change. It started with 'orderly' pronounced as 'urdely' by men, then was called batman, changed to sahayak and now has come to be called 'buddy'. What is in a name, a rose by any other name will smell as sweet. He is not a servant as some civilians want to believe. He is a proud soldier with a specific task. He has a special relationship with the officer, especially if he is a bachelor. They are close buddies with a good rapport. In battle, they rely on each other. The sahayak is required to take care of all mundane but important issues so that the officer as a commander can devote himself to the task of looking after his command, and accomplish the mission at hand without having to bother for himself. He is the officer's shadow, his buddy, protection and a runner in battle. Of course, he also gives him his morning mug of tea and readies his tent and uniform.

I remember we were committed in intense counter insurgency operations in the valley from 1992 to 1994. My sahayak cum buddy, Naik Mahesh Singh, a wonderful Kumaoni soldier had been with me for a few years. He would always accompany me on operations and was my personal protection. Those were the days when there were no bullet proof jackets and we operated in shirt cellular OG against the terrorists' AK 47 bullets. The men had modified pouches which carried the spare magazines and that served as some protection against terrorist fire. One day, operating in Sopore where our battalion was deployed, we came under intense fire in the built up area. Mahesh quickly moved and covered me with his body. We moved behind cover and then I asked him, "Like me, you too have a family and you have three daughters. I notice every time there is danger, you cover me with your body and you have no more protection than I have." He replied ,"Sir, my job is to protect you. I will never be able to live in peace if something happens to you without me coming in the way." Wonderful spirit, the sahayak is not what some civilians portray them to be. They are officers' right-hand man in battle.

Well in peace time, it is a little different, even though his task remains the same. A bachelor has a simple life style, he has nothing to hide or seek. He could be sleeping in his frenchy underwear without a worry of who is coming or going. So the sahayak has full freedom to roam around the officer's room, irrespective of what is happening. The sahayaks of bachelors normally remain with the officers for long durations since they spend very little time of the day on the officers' requirement. They attend all aspects of training and unit activities which keeps them professionally sharp. As I write this, Havaldar Ganesh Bahadur who was my first sahayak in 1979, gave me a call the other day just to say hello; more than thirty two years later. So you know what our relationships with our buddies are.

Banta Won't Let you Down

Captain Sandhu from Punjab had a sahayak - Sepoy Banta Singh, also from Punjab who had been with him for long. Banta was used to his no holds barred bachelor ways. Sandhu got married and landed with his wife to stay in the same bachelor's room. Banta could not visualize that things had changed and he should be more careful entering the room now. He would just walk in surprising the lady or the couple. Sandy explained to Banta that now he should not barge in without ringing the bell. Despite the advice, Banta would land up with morning tea and suddenly appear in the bed room . Sandhu could not be too harsh with him, for he had served him well for years, in peace and in combat. Sandhu again told him,

"Bante thora dhyan rakha kar. Ab toh meri shaadi ho gayee hai, mem saab bhee hain, jara ghanti baja kar andar aya kar."

(Banta, now I am married and my wife also stays with me. You should not come in without ringing the bell). All this while Sandhu had considered himself lucky that despite Banta's surprise barging in, he or his newly wedded wife had not been caught off guard showing each other affection.

But Banta Singh would still do the same, push the door without knocking and if the door was not bolted, he would land up in the room with tea.

Sandhu yet again advised Banta that he should not enter the room without knocking if he found the door open.

Banta replied *"Saab tussi phikar na karo, ap ko koi problem nahi hogi, mein andar aun toh pahle darwaje ke key hole se dekh lenda ki andar jana chahida ki nahin."*

(Saab, you do not worry. You will have no problems with me. Before entering your room, I peep through the key hole and check whether I should go in or not).

Sandhu realized only then why he or his wife had not been caught off guard so far.

Man Proposes God Disposes

Unemployment remains a major problem in India. For those who are high school pass, army remains the best option for a life of honour and dignity. I was posted in the Regimental centre in 1998. Centre used to conduct recruitment rallies periodically. The soldiers posted there would request the officers to help their brothers and cousins get recruited. I had a sahayak from Bihar-Guardsman Jha. He came to me with the same request. I said, I cannot ask the presiding officer to do this, there is a fair system. Your brother has to pass on his own merit. I can help you to ensure that he gets a fair chance.

Jha said, "Saab, he has already passed the test. You only help me in his medical. The required weight is minimum 50 kilos but my brother is 49. That is nothing, he will gain one kilo in no time during training."

The logic was right, but the doctors have strict rules to follow and thus I was not very keen to ask him to do so. Nevertheless, I said one kilo was easy to manage. I suggested that he must feed his brother one dozen bananas before the medical and that should do the trick. He liked the idea. In my house there were a dozen bananas lying. He weighed them and it was perfect with adequate room to spare. I relieved him the next day so he could help his brother.

He returned home that evening and I could sense that he was not too bright. I enquired about his brother's recruitment. He said his brother could not make it in the weight.

"What happened, did you not give him a dozen bananas?" I asked.

"Yes I did," he said.

"Then what happened?"

"What do I tell you saab, my fate is such. He ate a dozen bananas, his weight became 50.5 kilos and I was very happy."

"Then why didn't the doctor pass him?" I asked him.

"Not his fault saab. There was a long queue for medical. By the time my brother's turn came, he could not hold and had to ease himself. And before I could get more bananas, the doctor had taken his weight. He was back to 49 kilos and failed."

Well, it was literally a case of a dozen bananas down the drain. I felt sorry for him.

Passing Orders...Murphy's Law

Communication skills are important in any profession. In the army the repercussions of misunderstandings can be devastating in war and peace. The orders can really get distorted in the army as they are passed down the command channel.

As a young officer, I was advised by my commanding officer (CO), Lieutenant General Pup Mann (Retired), "Never pass orders on the move i.e. while walking. Stand and pass orders deliberately." We also need to understand that the troops' exposure is different and thus need to be told very clearly what is required.

A battalion had its troops from Haryana who are generally non drinkers, and vegetarians, unless they can get it ready made on a platter. Their knowledge of liquor rarely goes beyond XXX rum. A mention of a Haryanawi soldier, referred to as Texan in the army parlance, normally conjures up the image of a rustic hardy soldier with a hard hitting Haryanawi accent, devoid of courtesies. In a battalion which had troops from this state, the CO was expecting some friends for dinner. He told his batman to go to the mess and bring a bottle of 'Peter Scot' whisky. The batman came back in time alright but with a patromax (kind of lantern that burns on kerosene oil).

In another incident the batman told to bring a pine apple brought a 'pani palla' (Hindi name for military rain coat).

The CO of a Maratha battalion, posted in Kupwara, Kashmir, in late eighties was in for a surprise. The head clerk reminded him that it was his wedding anniversary that day. Good soldier that he was, he did not remember the all important day. First thing, he called up his wife and earned some brownie points. Next, he chose to organize a small celebration with his officers. Realizing that nothing fancy could have been done in that remote place, he scribbled something on a note pad and asked the subedar major to get two dozens of those by afternoon.

Afternoon came and went, the stuff was nowhere to be seen. The CO wondered why the hell was a simple thing taking so long. The subedar major told him that it was not available nearby and a man has been sent to Srinagar. By then the CO had lost interest in his wedding anniversary.

Well, finally the man arrived with the stuff in the evening. He had brought two dozen *'panties'.* The CO had scribbled two dozen *'pastries'.* How does one throw a wedding anniversary party with two dozen panties in a field area?

I had a very sincere sahayak from Assam. Commanding my battalion then, I had invited some guests. The wonderful sahayak had helped smarten up the place and spruce up the plants. Later in the evening, I was impressed to see that the leaves of my croton plants were shining bright like the toes of drill boots. Damn good, I thought, how did you do that I asked him. He said, *"Saab tel maalish kiya hai."* He had polished them with mustard oil.

The damage was beyond economical repairs. No amount of 'fauri ilaz' worked. Over the next two days all my plants were dead for good. He was pardoned for error of judgment for his intentions were noble. (Not that I had any other option).

In a Bihar battalion serving with us at Udhampur in J&K, there were two officers by the name Captain Sucha Singh and Captain Buche Singh. There also is a field firing range nearby called Buchemandi. The adjutant passed an order through his JCO adjutant that Captain Sucha Singh with field firing party will go to Buchemandi.

From adjutant to the JCO adjutant, and from the JCO adjutant to the company CHM (Company Sergeant Major), the order was finally conveyed to the other officer that Captain Buche Singh with the Mess Havaldar (Officers' Mess Sergeant) will go to the sabji mandi (vegetable market). The officer was certainly amused. He had no business to be going to the vegetable market with the Mess Havaldar. When he asked the orderly NCO who had given the order, he replied *"Pata nahi saab, hukam upar se aya hai".* (I don't know, the order has come from the top and a similar answer from the next level). Quite often, it is difficult to trace the origin of an order. Ask anyone and you get the same reply; it has come from the top.

Ensure You Must...

The army trains you to do a thorough job, for, in the business of war fighting, there is no prize for the runners up. In doing so, we are trained to cater for all that can go wrong, in what is called the contingency planning. Yet, there are things that happen, which even with the best of training you cannot plan for. They say, when the troubles come, they do not come in singles-they come in battalions. And they will come when it hurts you most.

Inspections are an important part of military life. Detested though, they set things right, things which are otherwise forgotten. Where the visiting officer is staying overnight, his accommodation is one of the many important things into which a lot of effort and money is invested in. The Army Commander was visiting a divisional headquarters in Rajasthan. His wife was a known stickler for small things. No stone was left un-turned to make sure that the accommodation was just perfect in every possible way. Running water supply was erratic as it often is, so a syntax water tank was placed fully topped up. The window AC was checked many times over and run for long hours to double check its endurance. The opulent guest room looked and smelt great. Everything was just perfect.

The Army Commander and his wife stayed in the wonderful guest room readied painstakingly. The next day the lady was annoyed and fuming. She complained that the WC had no water, the AC had stopped working in the middle of the night, making it real miserable in the heat in that sealed room. To top it all, her brassier was missing. What kind of division was it that the Army Commander's wife's brassier could go missing. All the problems were fixed the next day under high level supervision. The staff court of inquiry revealed the following:

The WC was perfect, but before the Army Commander came, the guest room sahayak flushed the WCs to ensure that all the bathroom flushes were working. Then the guest room NCO checked them all. Then it was the officer in charge of the arrangements, followed by the second-in-command of the unit responsible, followed by his commanding officer, followed by the brigade commander and finally the divisional commander satisfied himself that the flushes were indeed working. A perfect AC was opened up to clean it up and reassembled. The military electrician who was pressed into service for the job could not put it together and left a gaping hole between the AC body and the cover. It was later learnt that the man who actually attended to the AC was a vehicle mechanic as the right one had proceeded on leave.

Later in the night when the VIPs used the loo, the flush did not work because the water tank had run dry because of numerous checks, starting from the sahayak to the divisional commander. As for the AC and the brassier - the kajoos, badams and the fruits that were kept in the room for numerous rehearsals attracted rats who had made their abode above the beautiful false roofing. Having had their fill at night, they were in no mood to sleep and played the whole night on the roof. The naughtier rats dragged a piece of cloth into the AC through the gaping hole which jammed something in it. The AC stuttered and coughed and went silent. Most important, the hunt for the bra proved futile. The AC was replaced the next day. The faulty AC was sent to the market for immediate repairs. It was there that the missing bra was recovered from the AC. Everything had to fit into place on arrival of the Army Commander. Beware of Murphy's Law.

In one military station, a new guest room was constructed. The first guest was a three star General from Delhi. The host GOC, like a good soldier wanted to check out the new Jacuzzi installed in the bathroom. He asked the sahayak if he knew how to operate the high tech gizmo. He gave a blank look, saying he had been detailed an hour ago and had no idea. The GOC saw a few knobs, not quite figuring out what each was meant for. He asked the soldier to turn on the taps one by one. As he turned the first one on, there was a 'crack of burst fire' type of sound. Both the GOC and the sahayak ducked for cover. The nozzles of the water jets went flying hitting the glass walls on the opposite side. Close shave the GOC thought, he could not have been declared a battle casualty making the supreme sacrifice in a jacuzzi.

The nozzles were fixed, this time with the added precaution of quick fix. The guest came and stayed. The next day, as they sat in the GOC's house for lunch, he asked the guest lady if there was anything that was wanting in the guest room, they being the first occupants. She said, she had not noticed much as they came late, but last night the sahayak barged in the bed room when both of them were on the bed. Well, in that fancy guest room there was everything except a latch in the bedroom. It had been inspected by many in the chain but everyone headed to check if the flush was working. Well, she was worried alright. The three star General was still doing well.

My unit was located in Bhatinda. That is a wonderful station where the power outage is rare. The divisional commander (GOC), a Sikh officer, accompanied by his wife visited the unit for the annual inspection. They stayed in the small unit guest room. That evening was the Dinner Night. As the commanding officer (CO) came into the mess decked in his formal mess dress, his second- in- command gave him the good news.

When the GOC and his wife were getting ready for the Dinner Night, the lights went off. The emergency light that was always there could not be found. Someone had removed it from the guest room that day itself for some unknown reason. The generator would not start as it had not been used for long, because the power had not failed in months. When the generator finally started, the electric cables in the guest room caught fire. The guests could not even run for their lives. In a room full of smoke and burning cables, the high profile couple was in a Catch 22 situation. The lady was half draped in a saree with her hair in curlers and the General half draped in his turban with the remaining three metres flowing on the floor.

The Gurkha sahayak was running all over the bed with a towel trying to put off the fire as the high profile couple watched with an indescribable expression. By some Godly intervention the fire was brought under control.

When the evening commenced, the ADC told the CO that the GOC wanted soup served at nine. The CO told the GOC "Sir, I understand that you have all the reasons to be annoyed after all the fiasco. Soup would be served as you desire, but as long as the GOC enjoyed the evening with the excellent jazz and pipe bands in attendance, we should continue." The GOC replied, "Not to worry, I am used to such things. Alright, we will go on as long as you have a drink for every drink I have."

The soup was served at half past two in the morning. By then two bottles of whiskey had gone down the drain between the GOC and the CO. The dinner finished at four. The GOC cut short his inspection, saying, a battalion that can handle crisis well will always be fit for war. The case of towering inferno was forgotten. (The CO in question was the author himself).

Military Justice

Way back in the early seventies, 3 GUARDS had an enterprising commanding officer (CO). A man of action and regimental to the core, his guidelines were clear. The youngsters couldn't have asked for more. One Saturday the youngsters of the battalion decided to paint the town red. They landed at the club and had a good binge. When they had finished, the bar lay shattered, the crockery was gone and a few pieces of furniture lay in many pieces. The civilian barman survived their wrath with a few broken teeth.

Promptly a report came from the station commander, giving the names of miscreants with a property damage bill, amounting to about three thousand rupees- a large sum those days. He directed that the damage be made up by the unit and disciplinary action be taken against the officers under intimation to the station commander.

The CO called a conference of all officers. The offending officers were dead scared.

"Welcome to the conference gentlemen," he said. He read out the report from the station commander. "This is the first report since we arrived here. These three young men have announced the arrival of the battalion in the station. For the first time since we have come, the town knows that the Guardsmen are here. The total damage is Rs. 3,000/- equivalent of four officers' monthly pay," he said. No one knew what was coming next.

"There are a few things that worry me as your tiger. We have eight subalterns in the unit, and yesterday only three were involved. Gentlemen we have to fight together in war, and unless we fight together in peace, we will not succeed in war. Remember, the Guardsmen live in packs and they must hunt in packs. As for the damage, it will be shared by the bachelors who were absent from the action. Second-in-Command, these three monkeys can proceed on fifteen days leave."

Justice delivered, no questions asked. What he told the station commander no one knows, one can easily guess though.

We Will Pray For You

One very hard working officer fell rather seriously ill, literally fighting for survival . In the unit he would handle almost three officers work load himself. The officers of the battalion would look him up in the hospital regularly taking home-made food etc. A day before he was to be operated, some said they would pray for him, and prayed for him alright in the unit mandir. The officer recovered and was deeply touched by their concern and camaraderie.

Discharged and back in the unit, he spoke to the officers and said, "I really appreciate your concern and am deeply touched. And of all the things, the fact that you all prayed for me, I have no words to express my gratitude."

One officer replied with a serious face, "We prayed for you only because if you had died who would have done our work. And worse, we would have had to do your share of work man."

Inter Caste Marriage and In Law's Feet

Captain Balbir is a fine Gurkha officer. Serving in the North-Eastern part of the country, he met a charming Christian lady from Nagaland and soon they were married. As it happens sometime, the conservative families do not accept inter caste marriages. Balbir's parents did not accept the marriage to begin with.

As time passed, Balbir thought it was time to address the issue and decided to land up with his wife at his parents' place.

He told his wife, "The customs in your part of the country are different. According to our customs the daughter-in-law touches the feet of in laws. This is the first time you are going to meet my parents, don't forget to touch their feet."

"Which foot should I touch first, left or right?" asked his wife nervously.

Golf The Army Way

Golf is a great game played in great spirit in the army. Some people feel it is the way to go up in life. Others feel that if you have gone up in life, then it is a must to play. When you have a senior of value in your four-ball, his every shot is a great shot. His every bad shot is bad luck. His simplest of missed putts are because of bad greens or some caddy moving and so on. The golf courses have acquired a new name these days, 'environmental and oxygen parks'. There are many jokes going around in the army, not that they happen actually. Its fun nevertheless as we shall see.

The Wild Golf Swing

This beautiful game of golf evokes mixed feelings in the army. The golf buffs spend a lot of their time thinking about it, discussing the previous round or fixing up the four-ball for the day. Those who do not, justify how it is an old man's game, a waste of time and man power. The other ambitious variety starts playing to get into the right circles, a misperceived path to higher ranks. The game however provides its share of fun.

I was posted as Aide-de-Camp way back in the early eighties. One day we fixed up a three-ball for a round of golf. Other than me, there was Major Parthy, an Engineer officer and Major Paramjeet, PJ for short, a short Sikh officer. We decided not to take a mulligan (second shot) on the tee off. Contrary to the norm, we started with the high handicap first, which brought PJ to the tee. PJ, extremely well dressed with a bright red shade on his black turban climbed up the tee with the swagger of a new golfer, took some wild practice swings and addressed the ball. He concentrated hard on the ball for what seemed to be an eternity.

Finally, the much awaited moment arrived, and he started his ultra slow motion backswing. It was so slow that you had to concentrate to catch the motion of the club. After a long painful journey, the club head finally reached the top of the back swing. After some mental appreciation inside an apparently confused mind, he let go the mighty forward swing with full might at his disposal. When he had finished, he had turned full ninety degrees, and in real copy book teaching, was facing the green looking for the ball.

Where was the damn ball? Right where it was, still perched elegantly on the white tee, seemingly teasing PJ. His wild swing that could have dropped a Rhino dead had missed the ball by miles. He stood there gazing at the ball, red faced, with an expression Shakespeare would have found difficult to describe.

What next? There was no mulligan allowed, the ball was on the tee and there were two golfers still to tee off. To tide over the crisis he was permitted to take the mulligan. This time, more careful than before, he took even longer, and a more painfully slower back swing. The club seemed to be stationary at the top for a long time as PJ contemplated the down swing. He, then holding his breath, and keeping his head down, slammed harder at the ball, probably to make up for the lost shot. Massive shot it was alright. The club head hit the ground about a foot behind the ball, kicking up loads of sand and grass in the air, leaving behind a sizeable crater. The sand hit the ball which rolled two feet off the plastic tee. Clearly embarrassed yet again, he looked at us as the ball came to a stop in the centre of the tee box.

Major Parthy, who was always looking for fun promptly said, "That was much better PJ."

PJ was clearly flustered and took his third shot off the tee box as we did not want any more craters on the tee.

Woodsworth English

In 1984, with the 'Operation Blue Star', the military action against terrorists in Punjab over, life had settled down to a predictable routine.

We had started playing golf at Bhatinda where our division was stationed. The Colonel General Staff (Col GS) was a short stocky Sikh officer with a large turban on a large round face with big black round specks. He had learnt golf after taking over as Colonel GS. Being a good sportsman, he had surprisingly learnt the game well in a short time, a bit erratic though. For four consecutive days, the Colonel GS had been partnering the General Officer Commanding (GOC) while I partnered the Deputy GOC, also a Sikh, not particularly on good terms with English. On four consecutive days Colonel GS partnering the GOC had lost ten bucks to us. Forty bucks and the caddie fee coupled with the lost balls were discouraging in the early eighties. The monetary loss, it appears, must have been bothering the inimitable Colonel GS, more so, because he was partnering the GOC. The next day the GOC did not turn up for some reasons. We found the fourth ball in Colonel Puri, 12 GUARDS, son of film actor late Madan Puri. As we were walking up to the first tee, the Colonel GS, putting his hands on my shoulder said aloud for others to hear, "Come on partner, we are going to sort these guys out today." Normally the partners are selected by throwing the balls. The two balls lying closest become the partners. Colonel Puri said, "No way. That is not the way partners are decided, we will throw the balls."

The Colonel GS replied, "No way friend, it is army and in the army the senior can do no wrong, so Sengar is my partner, no arguments, period." The tussle to take me as their partner was because I was playing good golf those days. So it stayed, me and the Colonel GS on one side, and Deputy GOC and Colonel Puri became partners.

As the game progressed we were in trouble. As the luck would have it, I was having a terrible game and nothing seemed to be going right for me. My best efforts to come back in the game were in vain. The harder I tried, the worse it got. The Colonel GS who had picked me as his partner with so much confidence was fighting a lonely and losing battle, and of course cursing me. Finally, everything depended on the last hole. With me out of the game with two OBs (out of bounds with two penalty strokes), my partner, struggling very hard from the 'Rough', somehow won the hole by the skin of the teeth and squared up the game, saving a lot of money but still losing ten bucks.

As we walked back to the Golf hut, he said, *"Oh mundya,*

main lar ke tenu apna partner banata, socha thoda paise banayege, par tu aaj lid kar ditta." (Kiddo, I fought to make you my partner. I thought we will recover some lost money. But you were all shit today).

And then continuing in English spoke aloud, *"Never mind young man. Let me thank you all gentlemen for teaching me this aristocratic game of golf."*

The Deputy GOC who was walking right behind burst out laughing and said aloud, *"Ha...ha...ha, just look at GS Coronel speaking Woodsworth Angrezi."* (Angrezi- Hindi word for English).

Balls all...

Colonel Puri (son of actor Madan Puri), a Guardsman, posted in a divisional headquarter in Kota was a keen golfer with a good sense of humour. He was appointed temporary Golf Secretary. The golf course was a crisscrossing nine hole course in an area too small to accommodate two soccer fields, with a stadium and a Basketball court thrown in the middle.

Golf balls were not easily available then and pretty expensive. The division canteen (CSD) had golf balls that were strictly controlled. He wrote a minute sheet stating that the canteen has golf balls that have probably outlived their shelf life. Further, the balls in question are 1.64 inches in diameter, whereas the balls permitted in PGA tournaments are 1.66 inches. It is therefore recommended that these balls be sold to officers at subsidized rates. Put up for decision of the General Officer Commanding please.

General Awasty, another Guardsman and a perfectly honest man wrote on the minute sheet.

"Neither do I have any intention of holding a PGA tournament in this handkerchief sized golf course nor is our CSD supplying golf balls for PGA tournaments. Balls to be sold as per existing rates, notwithstanding the size."

Blisters Despite The Gloves

Ambala cantonment has a good golf course. Brigadier Menon from Engineers was the Captain of the Course. He was always present whenever there was a competition, any other function or to monitor the upkeep of the course, but would not play golf himself.

I told him that he was there every second day, but what a pity, he was not utilizing the opportunity to learn golf and simply kept waiting. He took my advice and promised to start. A few days later I saw him again at the course during a medal round, well turned out, standing and waiting with an impressive bowler hat. I asked him,

"Not started playing yet?"

"No, I have," he said.

"Not practicing today?" I asked.

"No yaar I did, but I got blisters," he said, showing his left hand to me.

"Blisters on the left hand! Did you not use golf gloves?" I asked.

"I did yaar, but that was on the right hand na, right hand is okay."

I wondered how he had managed to put on a left hand golf glove on the right hand. He was just the right man to be the Course Captain of an Army Golf Course-innovative.

Brig KH Singh, now a General, is a thorough gentleman and a fine soldier. Two courses senior to me in the same squadron, he had that wonderful reputation as a cadet too in the NDA. He was commanding a brigade in the corps where I was posted on staff as a colonel. He stayed over for the weekend at Ambala. The next day I went for a game of golf and was pleasantly surprised to see him there with the Golf Secretary, all set for the round. I was asked to join him for the round. As I stood chatting, a soldier walked up with a tray in his hand with the Kharga golf cap, a packet of tees and two balls. The secretary presented him the cap and the balls on behalf of the corps commander.

"Well, do you notice that he has presented the balls only to you and not me, why so, may I ask?" I said.

"Simple," he said, "because by the time you reach this rank, you have no balls left."

We play
100-100-200,
any doubts?
??
??
ARMY GOLF COURSE
VIP VISIT : RULES OF
ENGAGEMENT
• PUTTS WILL BE
GIVEN AS UNDER :
★ - ONE CLUB LENGTH
★★ - TWO CLUB LENGTHS
★★★ - THREE CLUB LENGTHS
★★★★ - AFTER TEE OFF

Fairway is
a neatly
trimmed
strip of
land that
separates
Golfers
searching
their balls
Sir, you hit the
fairway regularly.
How do you
manage that?
I always aim
at the jungle.

Never seen you take a practice swing, sir?
I have very few good swings, don't want to waste one on a practice swing.

Sir, he is giving balls only to you, why?
Because by the time you reach this rank, you have none left?

There is Not to Reason Why.........

The incident below is yet another interesting story. It demonstrates in its own way, funny though, what makes the army an organization that delivers by virtue of military discipline and blind obedience.

The army works on two rules.

Rule 1- the senior is never wrong.

Rule 2- if the senior is wrong, see Rule 1.

The 'senior can do no wrong syndrome' has taken deep roots in the Indian Army. Stupid it might appear at times, but the Kargil War could not have been won but for strict discipline and blind obedience.

Captain Dallal, 'a happy go lucky' bloke was the adjutant of a Punjab battalion on the 'Line of Control' in Jammu & Kashmir. Gifted with blood shot red eyes and a muscular build, he looked drunk twenty four seven-any time villain looks. He was called by his commanding officer (CO) and told that Brigadier Rama Rajan, from the Corps headquarters would be coming to the battalion for area familiarization and would be having lunch in their mess.

No problem said Dallal, the Brigadier was his instructor in the Infantry School during Young Officers' course and he knew him. His CO added, though he did not understand why, he was told to keep an inflated scooter tube handy for the Brigadier in the battalion. Dallal like a good adjutant managed two inflated scooter tubes from a scooter repair shop close by-one plus one reserve.

Dallal and his boss waited for Brigadier Rajan discussing what the tube was all about, failing to find an answer. "Surely he couldn't be having plans to float across Manawar Tawi into Pakistan," thought Dallal.

Two military vehicles steamed in raising all the dust in their wake. Out came Brigadier Rajan with an inflated scooter tube in hand. He met the CO and met Dallal warmly recognizing him as his old student. Dallal, a simple Texan had an inflated tube in his hand and kept wondering the name of the game. Fifty yards behind, hiding behind a bush was a Sikh jawan, waiting with the reserve tube ready to appear on a pre-decided field signal. Brigadier Rajan then handed over the tube to his driver. Dallal followed suit and handed his over to his runner. The team was then conducted around the battalion defences and brought to the mess for lunch.

As Ram Rajan approached the mess, his driver came running and handed over the tube back to him. Dallal followed suit and took his tube from his runner, wondering what he would do with it inside the mess. Brigadier Rajan entered the mess, placed the tube on the sofa and sat down on it.

Here on, the episode is reproduced verbatim as narrated by Dallal himself.

Quote, "Seeing him sit on the tube on the sofa, I thought it was some new custom of sitting in the mess, which we had not come to know being in the field. After all he was a Brigadier, he knows what he is doing. So I also placed my tube on the sofa and sat down on it.

Seeing me do that, Brigadier Rajan said, *"Oh! I didn't know you also had piles Dallal."* As he said that, my CO and other unit officers looked at me, rather embarrassed at my plight. Then I realized why he was going around with a tube. I sat there quietly; embarrassed, sheepish and shattered not knowing what to do. All the officers were served beer. When my beer came, much less drink, I wanted to drown myself in it for my stupidity in following him blindly. It took me many days to overcome the embarrassment." Unquote. So life can be funny and silly all at the same time.

Baisakhi and Good Friday

India is a land of festivals. The troops in the army come from all castes, religions and regions. The army has a system of giving restricted holidays to the units depending on the unit's class composition and the festivals they celebrate. It therefore happens that in a station you could have one unit having a holiday whilst the others are on duty.

Sometime in mid eighties, a Sikh battalion was preparing for the Baisakhi holiday (a primarily Sikh holiday celebrating a good harvest). The day arrived and it happened to be on a Sunday. Well, that was a bit unfair, the troops thought, losing a holiday. One spirited and wise surd chipped in – in Punjabi, *"Toh kee lore hai. Is ton pahle woh Naga battalion da Good Friday bhee Sunday ko seega."* {Don't worry yaar, it happens some time. Last year the Naga Battalion (Christian troops) had their Good Friday also on a Sunday}.

Recalling Beant From NDA

Well, remember we had talked of Beant's incredible experience in the NDA, having started each day of his first term with a stunning slap. He had finally managed to get through the first term alive and with passage of time had reached the Indian Military Academy. It was time to fill the choice of arms. That was always a big decision; it was a decision for life. He was prepared to join anything from Army Service Corps to Para Commandos. One place he never wanted to go was to the regiment which wore its lanyard round its neck, and that regiment, the only one in the army was the Maratha Light Infantry Regiment. In this regiment of a dozen and a half battalions, all but one battalion wore its lanyard around the neck, like a hangman's noose. For whatever reasons, Beant had that kink. Beant stood at attention in Chetwode Hall in front of the officer announcing allotment of arms, "Congratulations, Cadet Beant, you are allotted a fine regiment, Maratha Light Infantry. Wish you all the best." Beant did not know whether to laugh or cry at his destiny. Out of over two dozen Infantry regiments and many other options, he had to land up in the only place he did not want to be. He had never opted for Maratha Light Infantry. Unfortunately, man proposes and the army disposes.

He narrated this to me long ago, stuck for life with the lanyard around his neck. Incidentally, much later though, General JJ Singh, the Army Chief from Maratha Light Infantry changed that and the whole regiment started wearing the lanyard on the shoulder. That was almost a quarter century after Beant was commissioned.

Well, much later, it was 1984 that we were doing collective exercise in deserts. Beant landed there as the staff officer with his Colonel of the Regiment, who was the chief umpire on our formation. Beant being an old friend stayed with me in my small tent. I was surprised to see him come there with all the battle gear; the small and big packs, anklets, water bottle, helmet and all that rig. I pointed out; the exercise was ours but he had more battle gear than me. Beant replied,

"In the army the battle is for two things. First, being there in the right dress, and the second, being there at the right time. And remember, the secret of happiness and success in the army is- a junior officer must look like a 'phuddu' (clot) in front of a senior officer. This way, you do not attract attention. If you look too smart, he will get after you."

And then I said, "Listening to your logic, I bet your water bottle must be leaking." Beant burst out laughing. He said I was dead right, his bottle was indeed leaking. He said, with the chief umpire he will surely get water to drink anywhere, the bottle is only for the right turn out, and who cares if it is leaking.

Size Does Not Matter

The Sikh Light Infantry is a great infantry regiment that has great achievements to its credit in battle. The sturdy soldiers, all Sikhs, by and large hail from Punjab. They are very famous in the Indian Army for their resourcefulness. There is nothing that they cannot accomplish. Even for the most impossible task the JCO or the NCO would reply *"Fikar nahi saab, ho jauga."* (Don't worry sir, it shall be done). They are as capable of growing roses in sweltering deserts as capturing most intimidating mountains. You may find it interesting to know that their motto is in Latin and true to their spirit, it means 'Find a Way or Make One'. And they always do.

Interestingly, quite often, one problem resolved by them in their own manner creates many more. One battalion of this regiment was being relieved by another battalion in a peace station. On relief, a battalion gets involved in handing and taking over of buildings, station assets, some weapons and transport. This could become a nightmare. The military buildings maintained by the civil organization called Military Engineering Service (MES), are often in a poor state of maintenance. And when a battalion leaves the station, MES comes up with staggering barrack damages running into huge amount. The units have no way to pay such huge sums. The MES could make things very difficult unless it is 'looked after well'.

This battalion had a huge barrack damage bill against it and wanted the MES to regularize some part of it. The MES however would not budge. It was finally left upon the ingenuity of the troops to bring the MES to mother earth.

There was a MES bulldozer which was working in the battalion area. At night the bulldozer used to be parked without any guard, assuming that nothing could happen to the big brute machine and it could not be safer than in the battalion area. The next day when the MES workers arrived to work as usual, they were horrified to find the bulldozer missing. Immediately the matter was reported. There was great panic and confusion. Frantic hunt was launched to locate the machine. It is one damn big machine, it cannot vanish in thin air just like that, the MES officers thought. All efforts to locate the bulldozer failed. When all this was happening, this Sikh Li battalion maintained a stoic silence and did not bother to contact MES for their problems as they had been doing before. Knowing Sikh Li, someone in the MES suspected that this battalion had to be involved. The battalion flatly refused, saying neither they knew anything about it, nor are they responsible for MES dozer parked in their battalion area. The MES officers were in tears and at their wits end. The battalion simply seemed not to care about its barrack damages at all.

After much goading and pleading, this battalion told the MES that though they had nothing to do with it, they knew the whereabouts of the bulldozer. They offered to help if the MES agreed to write off their barrack damages. The MES agreed this time without a fuss.

"Now tell us where is the bull dozer?" asked the MES officer.

"Right under your feet," replied the JCO.

"What do you mean, under my feet, you must be joking," said the MES officer.

"No I am not," said the JCO seriously, "Dig ten feet under you and you will find it."

The MES staff stood with open mouth unable to believe their ears.

They dug ten feet down, shocked to see that the dozer lay there, wrapped in tarpaulin, precisely where it was left the night it disappeared. The battalion had dug the bulldozer down and covered it with earth overnight. They did not pay a single penny as barrack damage.

Whenever a Sikh Li battalion relieves another battalion, it is a test of their smartness. They can hand over the same vehicle twice with a straight face.

Chicken From the Line of March

What do you want to eat sir? Tea, carrots or chicken.
(Serves to know your men better than their mothers do)

These wonderful soldiers from the Sikh Light Infantry create a stir where ever they go. Good tough soldiers, they always spread caution amongst the other units around because of their reputation earned painstakingly over decades. Extremely resourceful, they can produce wonders, except that their officers may have to spend days to do the damage control after their action. No wonder they joke that a Sikh Li officer is more likely to get sacked before the rank of colonel. If he crosses that hurdle, then sky is the limit. General Ved Malik from Sikh Li was our chief a while ago.

The Regiment has many interesting incidents to their credit which only they can do.

During an exercise when a Sikh Li battalion walked through a village, nothing seemed to have happened. The troops served their young company commander steaming hot chicken for dinner. In the exercises, where you normally survive on basic 'Shakar Para' rations, chicken was a surprise luxury. On enquiring, he was told that while passing through the village they had just picked up a couple. "And where did you carry them?" asked the young company commander. "In the 84mm Rocket Launcher," replied the batman. These guys known for their resourcefulness can serve you steaming hot tea on a route march before you can say Jack Robinson. How, you may want to know.

Well, a kettle on a stove - stove in a bucket and the bucket in the hands of a jawan on the march. A mobile tea stall, shall we say. Great guys who look after their officers extremely well.

Watch Out They Are On Board

I had joined my battalion in Siliguri in July 79. Soon I was packed off to Sikkim where one company was permanently deployed. A Sikh Li battalion was deployed there too. A Sikh Li soldier, driving a jonga landed up with a flat tire. It can be a frustrating experience in a high altitude terrain, because hardly any vehicles ply in those areas whose help could be sought. The problems get compounded when one does not have a spare wheel, as this duo did not. The driver with the co-driver worked out a strategy to tide over the problem. Standing by the road side, some distance away from their flat vehicle, they thumbed a lift in a jonga which had no passengers at the back. It happened to be the vehicle of a battalion commander in which he was traveling alone. After some time they requested the commanding officer to stop the vehicle and got down saying that their destination had arrived.

The driver on reaching the unit carried out his customary check of the vehicle. Shocked to find the spare wheel missing, and worried that he may be asked to pay, he ran and reported to the adjutant. The officer gave him a mouthful for the lapse and wondered what had happened. When he learnt that they had given lift to a Sikh Li buddy pair, he was beginning to put two and two together.

En-route the Sikh Li pair had unscrewed the spare wheel of the commanding officer's jonga and dropped it on the road side. They then rolled the wheel down to their vehicle and got going. The driver drove to his unit and told the adjutant what he had done. The officer did not even look up from the file. He simply rang up the affected unit, who were much relieved and retrieved the wheel.

Courtesy at its Best

During a collective exercise a brigade commander visited one of his units, a Sikh Li battalion again. The drive back was long and would have got dark by the time he got to his headquarters. The battalion commander, realizing that, arranged an escort of four burly Sikhs to accompany the brigade commander.

After more than an hour of travelling, the commander decided to take a 'loo' break in a secluded place. As he appeared from the bush, the burly escort commander, in the best traditions of a good battalion, took the opportunity to ask him if he would like to have a cup of tea.

"Saabji, tussi chaa ya kofee kuch peena?" (Would you like to have a cup of tea or coffee?)

Tired from the long day and a bumpy ride, he welcomed the idea and chose to have a cup of tea. A good battalion, he thought, they have taken care of small things.

He asked for a field chair which his party was carrying and settled down expectantly waiting for a hot cup of tea. Fifteen minutes on there was no sign of tea. He wondered why it was taking so long as he presumed they would be carrying tea in a thermos flask. The escorts' body language indicated no tea related activity and appeared as if they were waiting for the commander to get up and move.

He called the escort commander and asked him,

"Kyon bhai, bahut der ho gai, tumahri chay ko kya ho gaya?"

(It has been a long time, what happened to your tea?)

"Saabji, chay da bandobast toh nahi seega."

(Saab ji, there is no arrangement for tea).

"Agar chay ka bandobast nahi hai toh koi gal nahi, par apne fir chay ke liye poocha kyon." (There is no problem if there is no arrangement for tea, but why did you ask then?) asked the

commander, a bit amused and a bit annoyed for wasting his time.

"Saab ji, tussi saade commander ho, saadi paltan which aye seege, sade guest ho. Poochana toh sadda farz banata hai na."

(Saab ji, you are our commander, you came to our battalion and you are our guest. Courtesy demands that at least we ask you).

Full points for courtesy, the commander thought. He picked up his field chair, got on the vehicle with a smile and resumed his bumpy ride after a fifteen minutes tea break sans tea. Wonder what he thought of the drills of the unit.

Why Does the Tank Have Forward Gears?

There is always a bit of friendly rivalry between various arms and services. The infantry and the armoured corps are the two major combat arms. The infantry sees itself as the arm that always delivers, no matter what, and there is no denying that. History is replete with examples, Kargil being the latest in the saga of courage and sacrifice by infantry. They tend to look at the armoured corps as a force that would have some excuse for not being where they should be. They would either remain in reserve forever or will not have the special brake fluid or engine oil or be chary of boggy ground or something like that (Not without reason though). Officers would joke that a single strand of barbed wire was enough to stop an armoured division on a minefield. That's not true anymore with all the kinds of equipment now available to ensure better mobility through the minefields. They pull each other's leg once in a while. Well, as far as infantry is concerned, its basic weapon was 7.62 mm rifle. Infantry's soldier's effective range thus is 300 yards. It lives a simple and austere life and relies for success on nothing but its men and morale and never fails. The others joke about infantry- *"Chotta kutar, choti soch"* (Small caliber–small thought). Well, with the 5.56 mm now in vogue, even that must have become smaller. The mechanized warfare is characterized by manoeuvre. An armoured officer while briefing on the map will spread his palm and say that he will be here. That palm covers half the map sheet, so he is fine if he is within ten kilometres radius of the target. Infantry officer on the other hand will say that he will be at that nala-track junction.

Both have concocted jokes about the other. One that is famous in the infantry about the armoured corps goes this way. I had narrated this to an American armoured officer and he was indeed amused.

"How many gears does a tank have?"

"Six, four forward and two reverse."

"Reverse gears are fine, they are withdrawing all the time but why the forward gears?"

"That is just in case the enemy gets behind you, damn it."

And then there is the air defence artillery, now called the Army Air Defence. By virtue of their tasks they mostly serve in good places. We often joke that their peace postings are Delhi and Bombay and their field postings - Chandigarh and Ambala. Because they rarely see field and high altitude areas, officers joke that if an air defence artillery officer was allotted a fourth floor accommodation in Delhi, he would start claiming high altitude allowance.They also have a very colourful master slide of air threat envelope that is used in every presentation from a Captain to a General. We pull their leg that an Air Defence officer only needs to master one slide to become a General. And so it goes for other arms and services too. As for these arms and services, they are all a proud lot and their mottos are simply inspiring. One ASC battalion calls itself 'Fighting 556'!

No one wants them to fight though, just drive safely and supply quality provisions. 102 Artillery or Engineer Regiment would possibly call itself 'Second to None' and The Guards- 'Pahla Hamesha Pahla' etc.

Just in Case I Have to....

The Indian Army primarily consists of Infantry. With huge mountain ranges bordering our neighbours, the Infantry operates on foot and has to lug a lot of weight on the back all the time. The soldiers land up carrying as much as thirty five kilos on their back in battle gear. The infantry soldier therefore has never nurtured the hopes of ever having the convenience of a vehicle under his bottoms. It has become a habit with him to keep thinking how to man pack anything and everything with minimum inconvenience.

A Gurkha battalion was carrying out infantry-tank cooperation training with an armoured regiment. During the break, one Gurkha soldier was seen going around the tank, seemingly confused. The officer walked up to him and asked him what he was looking at. The soldier innocently replied, he was wondering if they were asked to man pack these tanks across the minefield, where all will it open up from.

A Street Without Joy

The subject of military campaign for the Staff College entrance examination in the year 1989 was the Vietnam War. The prescribed book for this campaign was titled 'Street Without Joy'. Major Gurinder of 5 Grenadiers preparing for the examination visited a unit library in search of this book. He asked the library in charge, an Education Havaldar, another surd, *"Library mein 'Street Without Joy' Kitab haigi?"* (Do you have the book 'Street Without Joy' in the library?)

He replied, *"Saab fauj mein aisi kitaban rakhan da hukam nahi hai, pichee bhee ek saab dhoond rahe the."* {There are orders not to keep such books (meaning pornographic) in the library. Earlier too one officer had come looking for the same book}.

I couldn't blame the library NCO at all. I had got married around that time and I too was to take the Staff College entrance examination the same year. My wife, coming from a civilian background told me later. Seeing me go around the shops looking for the book 'Street Without Joy', she was worried if our marriage was going fine. A day after I procured the book, I saw her quietly flipping through the pages and seemed quite relieved to see that it dealt with the Vietnam War.

I Can Do as Well But For...

One of my course mates had done very well in the academy and was pretty famous in the NDA amongst the officers who had passed out around the same time. He joined his battalion in Siliguri. Six months later Second Lieutenant Mohan was commissioned into neighbouring 8 JAT in the same brigade. During the annual inspection of a medical battalion, both these subalterns were detailed to conduct their

BPET tests. When all that finished and they were walking back to their battalions, Lieutenant Mohan told the other famous subaltern.

"Sir, honestly, you are a very famous man at this young service and you have got a bright future. I was thinking that if I do all that you do, I should also manage to reach at least one rank less than what you would achieve. As I see, there are only three qualities which you have, that I lack."

"And what are they?" asked the other subaltern.

"First, you have a good voice. When you say Savdhan (attention) in your battalion, 8 JAT across the road comes to Savdhan," said Mohan.

"No problem, that is not so important for a good career, and what is next?"

"The second, you have a good sense of humour."

"No problem again, it helps a bit in some ways.You don't have to be a joker to become a General in the army. And what is the last?"

"The third is intelligence. You are very intelligent and I am not so," said Mohan.

"Well, that is a definite problem. My friend, you can do without the first two qualities. You cannot hope to get very far without intelligence. It's serious because you believe in it yourself."

The bright officer pepped him, "There is no reason to lose hope. Some wise man, I think Einstein, said, *"Genius is one percent intelligence and ninety nine percent diligence."* If you are prepared to work hard you will get there. You should dream young man, and it is only when you dream that you will reach for the sky.

He wished Mohan good luck and a successful career as they parted their ways to respective battalions. Mohan left the army half way through his career for a greener pasture, where he could do well sans a good voice, sense of humour and intelligence.

Happy Anniversary

One officer invited his unit officers to his wedding anniversary party at home. The party got going, and before anyone realized, it was two in the morning with the dinner nowhere in sight. By then, a lot of liquor had gone down the urinal.

The officer who had thrown the party suddenly got hold of his wife and said,

"Come on darling, let us go, it is well past two and the kids are alone. It is time we got back. Thank you gentlemen for a lovely evening. We enjoyed every moment of it."

The wife stood dumb struck looking at his face wondering what was wrong with him. She then said, "What is wrong with you. It is our home and it is our wedding anniversary party. Where do you think you are going?"

"Oh bloody hell. No wonder, I have been wondering the whole evening why these curtains were looking so familiar," replied her tipsy husband.

You Can not Fool Even One Man All the Time

The deserts can be a bad experience for those who have a poor sense of night navigation. Losing direction and getting lost is a common affair. One Gurkha battalion was camping in the deserts for training. After dinner in the officers' mess, one company commander proceeded to his tent. The officer lost direction and went around in circles not knowing where he was. En-route he saw the sentry pacing up and down with a rifle. You don't take chances with a Gurkha sentry, particularly when you do not know the password at night. So, to avoid being challenged, he decided to pre-empt the sentry, and as he approached the sentry post, the officer shouted,

"Kancha kastow cha?" (How are you soldier?)

"Theek cha huzoor." (Very well sir).

"Dal bhaat khayo?" (Had your dinner?)

"Khayao huzoor." (Yes sir).

And the officer proceeded unchallenged. After some time, the officer having gone around without being able to locate his bunk landed at the same spot and again saw the same sentry. To announce his presence again, went through the same conversation with the sentry.

"Kancha kastow cha?" (How are you soldier?)

"Theek cha huzoor." (Very well sir).

"Dal bhaat khayo?" (Had your dinner?)

"Khayo huzoor." (Yes, I had sir).

The same drama took place a few more times with the same sentry. The Gurkha soldier realized early enough that the officer was lost and was going around in circles. Next time when the officer landed at the same spot and asked the sentry, *"Kancha kastow cha?"*, the sentry replied,

"Dhanyawad huzoor, mato jann ramro chhu, ra beete ko aadh ghanta ma dherai dal-bhat khai sake ko chhu. Abb aunus, tapai lai tapai ko tambu samma purai diun, no tarah maile dal bhat khai ka bimar parchha."

(Thank you saab, I am fighting fit. In the last half an hour, courtesy you, I have over eaten. Come now, I will take you to your tent before I fall sick of over eating), and guided the officer to his bunk which was less than thirty metres away.

Carpet , the Rolling Thunder

Resource crunch, thy name is army. Improvisation, taking out sand from shit is the way we learn to function.

Posted in the Guards Regimental Centre, Kamptee, we were celebrating our 'Golden Jubilee' in 1999. Many functions were organized over the course of four days.

Two long red runners were purchased for the occasion. They were to be lifted immediately after an event and leap-frogged to the next venue. A dedicated party under a NCO was detailed and rehearsed for each of those. They would tuck in- in a trench behind a bush in the vicinity of the venue. The moment the VIP left, they would roll up the runner like lightening, load it on a waiting vehicle, drive like crazy and lay it out at the second venue before the VIP got there. Normally there was enough time, so there was no need for a tearing hurry.

The passing out parade of the recruits was the first event of the day. As the last recruit marched out of the parade ground, Colonel of the Regiment turned and started walking off the dais. He had hardly moved half way over the runner to the car, that the 'Runner Task Force', on a signal from the NCO stormed from multiple directions where they were hiding. Oblivious of what was going on ahead, they got down to the task for which they had practiced hard-head down rolling the runner in perfect unison- less than fifteen yards behind the VIP- walking on the same runner. The crowd trailing the VIP prevented him from seeing the rolling thunder. A great sight it was.

The Army Mule

The army rules and regulations can be really interesting. One commodity which often is a source of amusement is the army mule. The army mule is a much privileged being in more than one way. In operations, court of inquiry may be dispensed with on the death of a soldier, but you have a lot of trouble on your hands if a mule under your jurisdiction gets hurt or injured. A mule has a standard scale of authorized rations like a soldier. In the past, he was also authorized rum like the soldiers are. He exactly knows what that means and would be a reluctant worker unless he gets his rations as per scale! He has a buddy soldier called a Mule driver who feeds him, cares for him, gives him regular massage and virtually lives with him throughout the day. They can't live without each other. There is a funny side to these mule drivers as well. Lot of these men, when enrolling in the army, believe they are being enrolled as drivers, as the name suggests, and that they will be driving some kind of vehicle called mule. Alas, it is not so, it is one long tough career walking up and down the hills maintaining posts in the difficult mountains.

Over a period of time he has got used to getting Sundays and holidays as off days. Even the new 'recruit mule' seems to have learnt that by instinct and knows all that, the day he arrives in the unit. Since these animals are mostly bred within the army, I guess it is programmed in their genes. They know by instinct how much load they are supposed to carry and will detest carrying any awkward load. They also know that they are not supposed to carry load downhill except in an emergency, and know which emergency deserves that kind of effort. He knows that he is entitled to a military funeral on departure to his heavenly abode. The armies probably have a better documentation system for the mules than for a soldier. While the authenticity of this story can't be established with certainty, this story has been making rounds in the army for a long time.

In a difficult terrain in the Eastern part of the country, one mule fell over the hill and died. There was much hue and cry over the incident. A court of inquiry was ordered. The presiding officer questioned a host of witnesses.The vet scanned his medical history and submitted his report.

The court thoroughly interrogated the mule drivers on the mule's behaviour in last twenty four hours. The mule driver went into a shock, on the death of his companion. The doctor declared him unfit to give a statement for four days. After much investigation and harassment, the presiding officer recorded the opinion of the court:

"The court is of the opinion that Mule GS Number 34782140A of 243 Animal Transport Company committed suicide by jumping over the hill on 24 Jul 83. No one is to blame. The loss be borne by the state"

The opinion of the court created more hue and cry than the death of the mule itself. The commander was infuriated at the opinion of the court and summoned the presiding officer. He however stuck to his guns. He said that as the presiding officer of the court, he has to conclude from the accounts of the witnesses.

There was only one eye witness and he explained that the mule for last few days was under some sort of depression. He was not keeping too well, not eating well and shunning company of other mules by remaining lonely and threatening to bite. To add to the mule's woes, the work load in the past few days had increased due to a rare patch of clear weather after a long time. He was a disciplined mule and always led the section convoy maintaining the right pace taking only the authorized breaks.

On the day of the accident, the mule was walking ahead of the mule convoy. Suddenly he increased his pace and created a gap between himself and rest of the convoy. The mule driver was hard pressed to keep pace with him. He then stood at the end of a precipice, looked around, looked at the sky, looked at his driver with an expression as if to say good bye, neighed once as if he was yelling the battle cry and jumped to his death. Efforts by the mule driver to pull him away from the precipice failed as he stood like a rock.

In the absence of any other evidence, the verdict of the court was accepted as the folklore goes. The mule units were thereafter warned to keep a tag of mules with suicidal tendencies and ensure that they're not left alone. Orders were passed that the mules will not be over taxed; jawans could be used instead! The mule drivers were put through intensive capsule of stress management for mules.

And do you know, the mule driver works harder than the the mules, but no one talks of his stress management. This military folklore is passed on and on.

Rifleman Ghasita Ram...

Major Chillar, a Texan (a Jat from Haryana), joined a battalion which had Jat troops. The officer had a tendency to stammer and sometimes uncontrollably. One day he was detailed to disburse the pay of another company. The pay disbursement is itself a kind of comic parade.

On one side sits the Company Havaldar Major (CHM) with an aide. He collects battalion fund subscription, right at the source, and notes who was collecting the pay of the absentees. In the centre of the corridor sits the gang of four, the company commander, the company clerk and the two JCOs. A runner is in attendance to keep the uninterrupted supply of tea and oily pakoras going. The soldiers line up as per a list called the 'Acquittance Roll', too complicated to be pronounced correctly by the soldiers, thus called 'accounts roll' for simplicity.

The pay parade started and a soldier walked in, smartly saluted and reported.

"No 13667731 Rifleman Ghasita Ram, Alfa company, No 2 Platoon pay parade ke lie aya hoon sri maan."(No 13667731 Rifleman Ghasita Ram, Alfa company, No 2 platoon reporting for pay parade sir).

He was handed out his measly pay of rupees five hundred after having been counted by the gang of four- four times over. Ghasita Ram stood aside, deliberately counted the five-hundred rupee bills many times over and reported all correct in parade ground fashion.

"No 13667731 Rifleman Ghasita Ram, paanch sau rupay theek paya sri maan." (Rifleman Ghasita Ram received five hundred rupees, all correct, sir. *'Ghasita' translated in English would read 'to drag').*

Chillar had never heard such a name before and found it funny. He said, *"Kya naam bataya, Gha....Gha....Gha..... Gha...sita Ram,"* stammering badly, struggling to complete the name. (What did you say your name was- Gha....Gha... Gha....sita Ram as he stammered?)

The jawan who had a good sense of humour replied, *"Haan saab, hai toh Ghasita Ram, lekin utna lamba nahi jitna aapne ghasita."* (Yes, my name is Ghasita Ram but not as long as you have dragged it).

A Wrestler and Map Reading

The army pays a lot of importance to sports. The units work very hard to win the formation sports championships. Some units would recruit good sportsmen and leave them completely free round the year to train. In the bargain the sportsmen tend to miss out on basic training and find themselves lacking in their promotion examination. The units on their part try to help these sportsmen to pass the promotion tests by the so called 'mel - milap', liaison with the examiners. Mostly that worked and the sportsmen would get through and get promoted.

A Jat battalion had an illiterate national level wrestler who all his life had done little else. He knew as much about soldiering, as I know about wrestling - nothing. The battalion approached the presiding officer of map reading test, on which depended his promotion to Havaldar. Since the board consists of a number of officers the presiding officer said that he could not pass him outright. He would however ask him a very simple question, to indicate River Jhelum on the map and on the ground and he will be passed.

No problem the battalion officer thought. The wrestler was put through extra coaching to hold the map right side up and identify river Jhelum on the map and ground. River Jhelum is one big river that meanders through the famous city of Srinagar in J&K. The whole night the wrestler practiced showing the river on the map and was ready for the show.

The next day he proceeded for the test. The presiding officer possibly forgot what he had discussed about him the other day, and instead of asking him to show River Jhelum, asked him to show the capital city of Srinagar. That was one big surprise for him. He had not prepared for this question. There was no way he could locate Srinagar city on the metre map. He had not seen a map before this episode and could not make out the difference between a town and a tree. He stared at the map for what seemed an eternity as the presiding officer waited patiently for the wrestler to crank his unused brain into motion. Not able to resolve the puzzle of locating Srinagar, he finally said,

"Mein boloon saab, mein toh saari raat Jhelum nadi kee tayari ki . Mein beroon, keee saab pooch rahe ho Srinagar kit sai. Srinagar ka achaar dalega ke? Mane poocho ke Jhelum kit se, toh mein yun lath gaar doon."

(The whole night I practiced to indicate River Jhelum on the map. Now you are asking me to indicate Srinagar town on the map. Are you going to make pickle from Srinagar? You ask me where is River Jhelum and then see how I plant my flag).

Texan and the Artillery Round

The Texans have their own inimitable style of talking, shooting from the hips, shall we say. Those not used to their style find it offensive, but they mean no harm. It is the Texan style. A new artillery gun, the Bofors were inducted in our army. It generated a lot of interest because of the longer range and better capability. Many senior officers would visit the firing of these guns on the ranges.

A General officer witnessing the firing asked a Texan JCO as to how far does the round of this gun go?

"Ghani door jawe saab," (Goes a long way), replied the JCO.

"Batao saab, iska gola kitnee kilometre door jata hai?" (Tell me how many kilometres does the round go?), asked the General wanting to know the precise range of the gun.

"Bataya na saab manne. Main gole pe baith ke toh kabhee gaya naheen, lekin sunate hai ghani door jawe," (I have never gone with the bomb, but I am told it goes a long way), replied the JCO nonchalantly. The General knew it was time to move on.

Drive Faster Damn it

During a collective exercise, the defence minister was visiting the troops. One unit commander was given the responsibility to organize reception for the defence minister. Two helipads, one main and alternative, a kilometre apart were prepared for him.

The commanding officer (CO) with his second-in-command were waiting at the helipad for the defence minister who was being accompanied by the army commander. To his surprise, the defence minister's helicopter headed towards the alternative helipad. The CO along with his second-in-command mounted their jeep, CO himself taking the wheel and sped towards the other helipad. Seeing that the helicopter was almost about to land and he still a long way off, said impatiently *"Aur tej chalao, aur tej chalao,"* (come on drive faster- drive faster), as if someone else was driving. Meanwhile the second-in-command, the co-driver, watched in amusement wondering if his CO was ok. He then reminded the CO that he was himself driving. The CO smiled embarrassingly and said, "Oh, I was just pepping myself up," stepped on the gas and sped towards the defence minister. Tension can make you do weird things, doesn't it?

A Puzzle Called MS Branch

The Army Headquarter (AHQ) has many branches. Of them all, the one that affects officers the most, interests them the most and intrigues them the most is the holy MS Branch. Many have cribs, for right or wrong reasons since it deals with promotions and postings. They have a difficult job doing career management of such a large number of officers. Many officers believe that if you ask for a posting to north India, they will send you to south, and vice versa, and not without reason. Many actually contemplate asking for south if they want to go north and vice versa. Brigadier Subrahmaniam told me that he always gets his posting close to his home town. I asked him how does he manage that. He said, "I am Subrahmaniam, (Subrahmaniams hail from the state of Tamil Nadu in south India), they always post me to Delhi thinking I am from Kerala. They do not know that we settled in Delhi over thirty years ago. My first priority in life is to maintain that state secret."

As for me, I was posted in a remote corner of Kashmir valley in a hot counter insurgency area. I rightfully requested for a posting to Delhi, where I had never been posted in thirty years. I was posted to Nagaland, the extreme eastern end of the country. One officer gave four choices of posting-Chandigarh, Calcutta, Ahmadabad and Bangalore. The MS Branch was confused with this choice of North-East-West-South as they found it difficult to post him. When asked by someone about the idea of his choices, he said he was not interested in any of these places. He was interested in going closer home somewhere in central India, and given the MS Branch approach, he hoped this trick will work. That is actually what happened. He was posted to Nagpur, his home town. His strategy and smart choices had out smarted the MS Branch. Many have similar experiences. Of them all, here is one that is pretty interesting on their logic.

Captain Jham, hailing from Rajasthan, was a wonderful officer commissioned into my unit. He suffered an accident and was transferred to Ordnance Corps. By the time he finished his courses, he had about four years service. Getting close to marriage, he thought he must finish the high altitude tenure so that after his marriage he could stay with his wife at a family station.

He requested the MS Branch for posting to Leh- a high altitude area in northern part of Jammu and Kashmir. Despite all the assurances, he found himself posted to Ambala. When he enquired, the MS Branch told him that Ambala is only forty five minutes away from Leh by air. So long MS Branch. We often joke in the Army, that MS Branch guarantees eighty percent satisfaction. If you ask for Madras, they are certain to post you to Dras.

The Commanding Officer Asked for it

Colonel Bishnoi, commanding an Infantry battalion was an irritable bloke, in the habit of ticking off his driver for minor things. It was happening rather frequently and the driver was fed up with his ticking off. One day when he was reversing the vehicle, the commanding officer ticked him off, saying “How do you drive the vehicle, how much have you driven in your life? Two thousand kilometres?” he asked.

The driver, another Texan, who had- had enough said, *“Kya bol rahe ho saab, itna toh main reverse gear mein chala chuka hoon.”* (What are you talking, I have driven more than this in the reverse gear alone). He was never ticked off again.

General Points

A Major who decided to speak on a sand model after the General had given his mind, "Gentlemen, you just heard the General make general points and now you will hear a Major make major points."

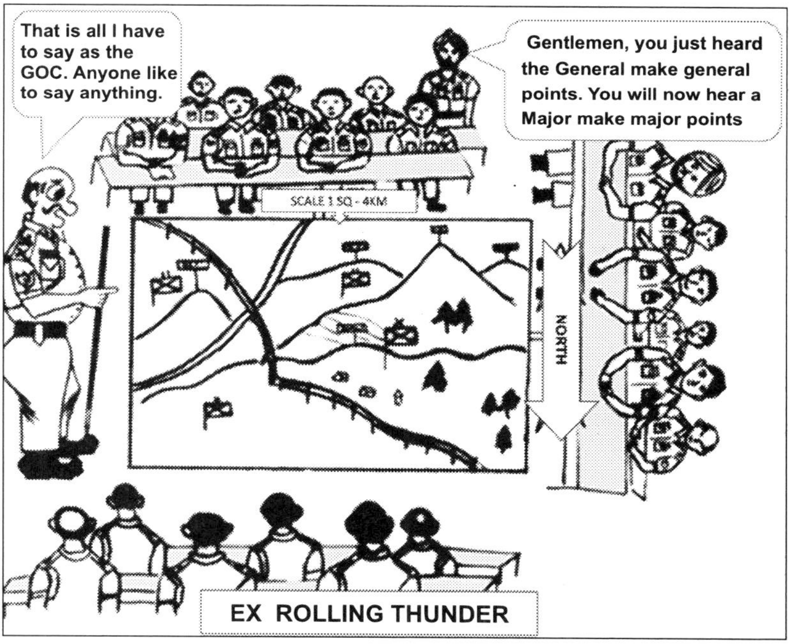

FROM THE FRONT LINES

Indian Army's history is saga of valour and sacrifice. Kargil is just the latest shining example of achieving the impossible in the highest battlefield where a war was ever fought. There are innumerable acts of supreme sacrifice and valour, one better than the other. Amongst them stands out the battle of Rezang La against the Chinese in November 1962 where 13 Kumaon fought a magnificent battle. Out of a company of 124 soldiers, 110 died fighting against all odds. Only 14, wounded in different ways, survived in the battle to defend an innocuous village, Chushul, in Ladakh. Freezing cold had frozen the dead in their battle positions, some with hand grenades in their hands. The snow had laid a shroud over the battlefield. Eight months later a shepherd wondered on to Rezang La when the truth of the battle was uncovered. Written on the memorial of Rezang La martyrs is this verse:

"How can a Man die Better than Facing Fearful Odds.
For the Ashes of His Father and the Temple of His Gods."

It is from such traditions of glory and sacrifice that the Indian Army derives it strength.

Indian Army is one of the most operationally experienced armies in the world. They have been constantly in battle since independence. Apart from the four major wars that it was forced to engage in since independence, it has been constantly committed in counter insurgency operations in different parts of the country. War movies look fancy and romantic on the silver screen, on ground it is far from so. There are no Rambos in real life. One bullet is all it takes.

It is a hard slog, mentally and physically demanding, through all kinds of hardships. Hungry, freezing cold and wet, out of sleep, far from dear ones, distant from an unaware public, slogging in the most hostile weather and terrain. It is a business that costs lives, lives of comrades that we have lived with for decades. It is the resilience of the soldiers that they weather the storms and get on with the job in hand.

The soldiers are a great lot. They will find a way to amuse and laugh at themselves irrespective of the hardships around. It is quite amazing how they can do so, when it is the question of survival.

Our army made a major contribution to the victory of the Allies against the Axis forces in Africa and Asia, both in the First and the Second World Wars. Many of their folklore are remembered even today, mostly relating to those wonderful Gurkha soldiers who have left their mark with their valour and courage.

Many politicians have the habit of meddling with the functioning of the army, in particular during war. In India Krishna Menon was one such character during the Indo-China war in 1962. Hitler of Germany was yet another, taking his country down the path of destruction. Winston Churchill, the famous British Prime Minister during the Second World War was also infamous for this trait. The Generals detested his meddling. He was known to directly ring up and talk to junior level office holders and then use that information to surprise the Generals. One day he rung up the war office in the middle of the night and enquired:

"How're the operations going on in Burma?"

"Great," was the reply.

"And how about the turning movement in North Africa?"

"Even better, I guess."

"Who is speaking?" asked the Prime Minister.

"Corporal Jones, duty corporal. Who is on the line?"

Churchill hung up without disclosing his identity, never to call up directly again.

Funny things happen in war. Indian Army was fighting in Burma in the Second World War. Those days majority of the officers were British with a handful Indian officers thrown in. Here are a few incidents from those days.

Sitting Pretty Amongst Confusion

The Second World War in Burma was raging wild. The Japanese attack created a panic and disorder amongst the troops. The artillery bombardment was sending up everything into smithereens and setting everything ablaze. The aircraft streaked in to strafe the area causing utter confusion as the men scrambled all over for cover.

Japanese were not the only foe; malaria and diarrhoea were no less, taking heavy toll. When all this was happening, a Second Lieutenant, badly down with diarrhoea, hardly able to control himself was looking desperately for a place to crap, running from place to place. He finally found a makeshift toilet. He barged in to the toilet, banged the door shut, pulled his pants down and shat. As he let go the salvo and relieved from his immediate worries, he looked up. On the bamboo door was written in chalk:

"In this moment of utter confusion, my friend, you are the only one who knows what he is doing."

TOILET
continued

My friend,
In this moment of utter confusion, you are the only one who knows what he is doing.
Bash on regardless
God bless !

Japanese in Delhi, No Way

Sepoy Sukhvinder was a young Sikh fighting the Japanese on the Indo- Burmese border during the Second World War. It was a long war and troops were sent on leave as best as the situation permitted. The Japanese were knocking on the doors of Kohima, Nagaland. Their aim was to get to Dimapur, the railhead which opened the gates to the plains of Assam and thence to the rest of India. Sukhvinder was granted fifteen days leave. He boarded the metre gauge train at Dimapur. Chugging along at snail's pace the train was painfully slow and stopped for eternity at every station. Sukhvinder was extremely disappointed that bulk of his leave was over by the time he reached Delhi, his home still two days away in Punjab.

At that time the British were having a hard time stopping the Japanese advance to Kohima. There was a rumour floating around that the Japanese were pushing the British hard and planning to reach Delhi in three months. When Sukhvinder learnt of this while on leave, his reply was typical – "Not if they come by the Indian railways."

Map; Your Best Navigation aid

It was during the Second World War in Burma that one Gurkha unit was ordered to withdraw under the pressure of Japanese attack. In the confusion of battle, one soldier lost contact with the battalion and got separated from the unit. He found a map and using it navigated for days through the jungles and rivers. Finally he reached the headquarters. He was taken to the adjutant where he reported that during his move back, he had marked the locations of the enemy units seen by him. The officer keen to obtain the latest enemy information told him to show the map. The Gurkha soldier took out a crumpled old map with enemy positions marked on it. The officer saw the map and was stunned. Incredible he thought, the Gurkha soldier was completely illiterate. He had navigated for a week in the mountains and jungles of Burma and marked the enemy positions on an English language map.

It was a London street map he had found somewhere.

Gurkhas are mountain people with a natural sense of direction in the mountains and jungles. They need no maps to find their way.

That's the Spirit

The Gurkhas have always been known for valour in battle. There are stories galore of their courage and sacrifice. When the British left India, they took along a few Gurkha battalions and the Indian Army of course has many regiments of the Gurkhas.

During the Second World War in Burma, there was a requirement of recruiting paratroopers from a Gurkha battalion. The commanding officer (CO) asked the company commander to ask for volunteers from his company. The company commander gathered the company and asked for volunteers to jump from the aircraft. After some hesitation and exchanging glances, about thirty out of his company of hundred volunteered. Little more pep talk got a few more. Efforts on part of the company commander to get any more volunteers failed. He went back to the CO and told him the story.

The battalion commander was surprised. Knowing the courageous Gurkhas, he had expected more volunteers. He asked the company commander what exactly he had said to the troops. The company commander replied, "I asked for jawans who were ready to jump from the aircraft, and only thirty volunteered sir."

The CO who knew his Gurkhas well said, "OK, now go and tell them that they are required to jump from the aircraft with parachutes, and explain what it means." The company commander did that and when he asked for volunteers, the whole company without exception volunteered.

Late Field Marshal Sam Manekshaw, the Indian Army Chief during the Indo-Pak War of 1971, himself from Gurkha Rifles used to say, "If a man says that he is not afraid in battle, he is either a liar or a Gurkha."

So long, brave soldiers.

The Captain General

During the early days of Pakistan sponsored terrorism in Kashmir, the terrorists would become self styled battalion commanders or district commanders etc. Their privilege - control of funds and carrying an AK 47-UBGL (Under Barrel Grenade Launcher) while the rank and file carried only the AK 47. The genesis of this practice possibly can be traced to an interesting episode post independence. Soon after the independence in 1947, in a surprise move, Pakistan infiltrated the tribal raiders supported by their army to annex Kashmir. While Pakistan was fully committed to this act, the whole thing was pretty disorganized in the initial stages. An enterprising fellow from the present day Pakistan Occupied Kashmir had appointed himself the Defence Minister of Kashmir. The leaders of the raiders, many of whom were ex-servicemen had grabbed the opportunity of quick promotions. Soon they went up the ladder moving from self styled Captain to Major to Colonel. Before anyone realized what was happening, anybody who was somebody was either a Major

General or a Lieutenant General. Two with more muscle power became self appointed Field Marshals in the Poonch sector alone.

The self styled defence minister was attempting to set up some kind of an Army Headquarters with different branches. His guide was a British field manual which he always carried in his pocket and referred frequently. Soon he learnt of the fast track promotions that the leaders in field had helped themselves to. The defence minister could not be left behind in this race. He searched around for an appropriate rank for himself but learnt to his dismay that Field Marshal was the ultimate rank in any army, and there were two already. After much research he appointed himself 'Captain General', a German army rank, which to his logic was superior since it had no relation to the British rank structure adopted by his self styled field commanders. How I wish it was true today, we would all be Field Marshals.

The Walking Ghost

Captain Bansi Lal had never believed in ghosts. He was the duty officer on a cold chilly night in Sikkim in high altitude.

He decided to go out and check out the duties at mid night in the killing weather. Walking up to the magazine (ammunition dump), he almost dropped dead as he saw a sleeping bag bouncing vertically up and down. Ghost, he thought, his heart pounding. Too afraid to move, he froze. Soon the sleeping bag moved across and shouted, *"Tham, kaun atta hai?-password,"* the sleeping bag challenged the officer. (Halt, who is there? Password)

Bansi Lal heaved a sigh of relief as he realized that the vertical sleeping bag was a soldier on duty. Thambi on duty, a soldier from Kerala, where temperatures rarely ever touch

23ºC, found high altitude cold too much to bear. His improvisation was ingenious. He cut holes at the bottom of the sleeping bag for his feet, got inside the sleeping bag, wore snow boots and stood on duty, in perfect comfort.

Combat : The Holy Helmet...

Poonch is an important small border town located on the Indo-Pak border in Jammu and Kashmir. Strategically important, it has been in the news since independence. The army has been defending it from external and internal threats since then.

Pakistan had launched its vicious proxy terrorist war in Kashmir in late eighties, pumping thousands of foreign terrorists into India and sabotaging local youths on religious grounds for jihad. Youth in large numbers were taken across the difficult border terrain for terrorist training. Poonch by virtue of its location was an attractive option to spread the arc of terrorism.

In 1991 my battalion was posted in Poonch sector. Pakistani Army had forcibly occupied a remote Indian village on the Line of Control and taken the able bodied men for recruitment as terrorists for the proxy war. Rest of the villagers had run away to the hinterland and reported to the army posts. All efforts to resolve the issue peacefully had failed. Finally, my battalion was ordered to launch an attack, capture our village and re-establish own control.

Well, helmet and waterproofs are two important military equipment for a soldier in combat. But in peace time soldiers tend to detest both because they are unwieldy and uncomfortable to wear, and therefore avoid carrying them if they could help it. In peace time the steel helmet is seen as a dead weight and literally a pain in the neck. It is common to hear a remark that your ability to think comes to a sudden stop the moment you put the helmet on. For the water proof, they joke that if it rains for one hour outside, it rains for two inside. We as officers insist that the soldiers wear helmets during the exercises and firing, knowing we get cursed every time we do that.

All preparations done, the battalion debouched from forward defences after dark. It was pouring mad. The men scrambled on all fours on the slippery mountain side. As the leading elements advanced, still well short of the forming up place for attack, the company came under heavy fire from a close range. The MMG's long bursts rattled the dark silence and the hand grenades shook the earth as they sent chunks of earth flying in the air.(Actually the fire came from own patrol base of another unit from where we were to launch our attack. Interestingly, we were being guided by their own guides, a team of one JCO and two NCOs. So much for the dangers from the fog of war). Every one hit the ground looking for a fold to find cover. Some took cover in a drain less than a foot deep as tracers whizzed past over their heads. The chaos was complete. The situation was brought under control with difficulty and attack resumed, but we had already taken a few casualties. Later that night the village was cleared of the enemy and we dug in to occupy defences based on gujjar huts. The next night Pakistani Army counter attacked. They suffered very heavy losses and asked for a truce to pick up their dead. An uneasy ceasefire came into being and a period of grave uncertainty and tension prevailed.

I was commanding the outpost with two companies deployed for the balance of the crisis.

One happy go lucky Assamese soldier was later narrating his experience, over a cup of tea :

"Jab bhee saab log peace time mein helmet pahan ne ko kahta tha, toh mein sochta tha is garmi men faltu ka panga

sar par laga rakha hai aur pareshan karta hai. Jab attack mein taratar goli aya toh jis ko jo khadda mila us mein kood gaya. Mein bhee naali mein let gaya aur mere upar doh aur let gaya. Goli bahut der chalta raha. Meine pure helmet mein chupne ka koshish kiya. Agar helmet seer par rakhta toh body bahar, goli pair ke paas lagta toh helmet pair ke upar rakhta, to sir bahar. Kis tarah jaan bacha sirf bhagwaan ko malum. Us samay mein socha, agar helmet charon taraf se char char inch aur bada hota toh mein pura us mein ghus jata. Tab main soch raha tha afsar saab log theek bolta tha yaar helmet pahno. Tabhi toh woh afsar hain, nahin toh ham bhi afsar na ban jate."

(Every time the officers forced us to wear the helmets in peace time, I had cursed. I considered it unnecessary headache and a bother for nothing. That day during the attack when we came under heavy fire from close range, everyone jumped into the nearest pit, notwithstanding how deep it was. I also jumped into a one foot deep drain and two others landed on top of me. When the bullets were flying over my head, I would cover my head with helmet and the body would get exposed. When the bullets kicked dust near my body, I moved the helmet down and my head would get exposed. I was thinking, 'O' God, only if the helmet was four inches wider from all sides, I would have tucked in whole of my body. That day I realized that the officers were right. That's why they are officers, otherwise I should have been an officer as well).

The operation over, two forces were facing each other separated by less than fifty yards at places. Whilst the company was fully alert, there was a sense of uneasy calm with a suspect ceasefire in place. As a company commander I would walk through the company defences very often. One day as I stood near a Paki bunker, a bulky Paki soldier came out. Seeing our slightly built Guardsman Bheja Ram, he stretched his arms and patted his thighs like a wrestler in a show of strength. Bheja Ram pointing at the Pakistani shouted "What are you trying to show me? Look at me." Saying this, he smartly did a hand spring and a couple of back flips on the bunker roof. It really impressed the Paki soldier. He shouted to his other colleagues who collected immediately on top of the bunker. The Paki soldier said, can you show that again please. Seeing them gather, Bheja Ram shouted *"Ustad sirf ek bar namuna deta hai, chelon ab tum kar sakte ho toh kar ke dikhao."* (Instructor gives the demonstration only once, my students, now do it if you can). The Paki soldiers watching the fun melted away faster than they had collected. So long, brave soldiers.

There is God up there for sure

We remained deployed in village Kirni in an eye ball to eye ball situation till the issue was formally resolved two months later. We were at a great tactical disadvantage dominated by Pakis from all sides. Even eight feet deep communication trenches left us exposed from one or the other direction. While there was cease fire prevailing, Pakis could hardly be trusted. The atmosphere was very tense and the troops were in a high state of alert. If they chose to attack, Pakis would have been over us in a few minutes being no more than fifty to seventy yards, looking down on us from three sides. Half the company used to be on duty at night with lots of ammunition prepared ready to fire, and the balance slept in the bunkers fully loaded.

I was sharing a big room with a dozen soldiers in a gujjar hut. While soldiers were all sleeping on the floor, my batman had managed a 'look London, talk Tokyo' kind of a cot for me. My Assamese sahayak, the same fellow with helmet experience narrated above, noticed that I had been sleeping without a pillow. He asked me if I would like one. I said I won't mind if he could manage one. "But where will you find one here?" I asked. His reply was typical *"Saab, kuch na kuch toh karega."* (Will do something or the other). True to his word, I was provided a makeshift pillow. It was hard and uneven. Under the circumstances it was good enough I thought and complimented him on his resourcefulness.

A number of flag meetings later, the situation improved. Based on mutual understanding it was decided to pull out and fall back to pre-crisis defence line. I directed that stock taking of all stores and ammunition be done before we leave the post. The magazine NCO reported that some ammunition was not traceable and he would like to lay them out and count them to avoid problems later. Sure, go ahead I said, not realizing what he was looking for.

"Ja le aa," (Go get them), he said to my wonderful sahayak who was standing by, as if waiting for directions.

He disappeared and a little later strode out from my room with a heavy sand bag he barely managed to carry. Nonchalantly he started taking out 2 Inch Mortar bombs, laying them out - side by side in perfect discipline. Counting loudly in his typical Assamese accent, one, two, three…., he counted twenty one of them and said,

"Major ikees liya tha, ikees lotta raha hoon. Apka aur mera hissab kittab barabar, ok." (Sergeant, I had taken twenty one and I am returning them all. It's all clear between you and me, ok).

This fellow had come out from my room with a sand bag full of mortar bombs. I was wondering what the hell were these bombs doing inside my room.

I asked him, "Where have you got these from?"

"Got from your room, saab," he replied.

"And what were they doing in my room?"

"This was your pillow," he replied with a smile.

I was amused to see that all the bombs used for my pillow were without safety caps and half of them without even the safety pins. I had slept on that explosive pillow for over two months. And my meals brought from the company cook house were heated on a stove right next to my pillow.

Since then I believe "If God is with you, no one can touch you, not even primed 2 Inch Mortar bombs in your pillow."

Lesson learnt, when you ask a soldier to improvise, you better check. It could give you more than a pain in the neck.

Recalling Glacier Days

Siachen glacier is the northern most part of India. Siachen means a 'land of roses' in the local lingo. Far from it, it is an unending glacier with not a blade of grass. It is cold, hostile and threatening with temperatures dipping forty degrees below zero. Add the howling wind chill factor and you have the perfect recipe for disaster, should you make one mistake or dare to challenge the mother nature. This notwithstanding, it is enchanting and majestic.

Pakistan's sinister designs on this Indian unoccupied territory had led to the deployment of Indian troops on the glacier in 1984. It has since remained the highest and the most hostile battlefield on earth.

Siachen glacier in the initial days of occupation was a

tough place. The constant enemy artillery shelling added to the problems of weather and terrain. An innocent one hour walk from one post to another could easily turn into a catastrophe if the weather turned foul.The party could be lost for days in those deadly temperatures. A fifty metre walk from one tent to another in hostile weather could turn fatal in those white-out conditions. There was no scope for bravado, if you tried that, it would be at your own peril.

The Cult of OP Baba

I served on Siachen in 1986 when the facilities were at best rudimentary and casualties from the enemy artillery and the nature's fury were high. Indian troops are very religious. The piety is intensified when the threat to life increases. And Siachen is one such place; not so much because of the enemy, but from the unforgiving mother nature. It was common to see troops pray before any activity. Gods could be seen in every tent or hut.

Sometime later started the interesting cult of OP Baba. Although the origin of this belief has many theories, the most accepted version goes like this.

Once, the Pakis, always up to mischief, attacked a forward post in the Northern Glacier. Due to unavoidable reasons the troops temporarily withdrew from the post. The Artillery Observation and Fire Controller (Arty OP) JCO, Naib Subedar Om Prakash, however, chose to stick on despite all the problems. He declared that if the enemy has to capture the post, it will have to be over his dead body. Next day, when there was lull, these troops returned to the post but there was no sign of the enemy. Naib Subedar Om Prakash was not to be seen anywhere nor was his dead body recovered. A few days later this martyr JCO apparently appeared in the dreams of the Siachen Base Commander. The JCO propounded that he had saved the post from falling into enemy hands, and will never allow any intruder to step into our territory. He added that his soul will be present everywhere on Siachen to take care of our troops. The Base Commander later constructed a temple at the base camp on the name of 'OP Baba'.

On the glacier, 'OP Baba ki jai', can be heard wherever troops move. All activities of soldiers start and end with report to OP Baba. Teams moving out from a post give report to OP Baba, as one gives report to a superior officer. The same drill is repeated at the destination and troops thank him for his blessings on reaching safely. Such is the faith that staunch meat eaters and rum drinkers also abstain from these habits.

The photographs of OP Baba is kept in the temple at the Siachen Base with a complete set of uniform including belt, beret, shoulder title, lanyards etc. Every team which inducts in to the glacier prays and donates to the OP Baba. You may wonder what this is all about in this scientific world. Whatever be the truth, when it is a matter of life and death, any faith that gives a sense of safety and helps you overcome anxiety and fear is good. The more of it you have, the safer you feel. And all this helps you perform our duty fearlessly. The great thing about the army is that even if an officer personally does not believe in such things, he respects the faith of the men he commands.

Nothing expresses this sentiment better than this story. A Jat battalion was stationed in Lucknow. There was a Texan MT driver from Haryana who invariably missed the Sunday mandir parade. When one officer asked him, he said he did not feel the need to go to mandir every Sunday.

Sometime later, the battalion was posted to Kashmir during the peak of insurgency where the biggest danger was from the roadside bombs called IEDs. The officer sitting in his military truck saw the photograph of Lord Hanuman prominently hanging on the dash board. Amused, the officer pointed that out to him and reminded him of Lucknow. The Texan coolly replied in typical Haryanawi, "That was Lucknow saab. There were no IEDs or flying bullets there. This is Kashmir. Here I need Bajrangbali's blessings". A very honest confession, and no shame in it.

Dheere – Dheere Bol...

My battalion was deployed on the main glacier in 1986. The living conditions were far from adequate. The battalion headquarters was then at Kumar and the companies were deployed at most inhospitable passes on the Saltoro range up to 20,000 feet. There were only three FGH or the Fibre Glass Huts; one was the MI Room, second the commanding officers' hut and the third for the men. I, the adjutant and Captain Tarun Kaul, the regimental medical officer were the shivering occupants of a makeshift tent, made from multiple layers of parachutes. The skid boards covered with multiple layers of parachutes made the side walls. It was the only accommodation for any officer who transited through from company posts, another few days trek away. Food was good, only there was no appetite. The terrain, where mountaineering expeditions went earlier, had become a play ground for the army. Butter in plenty, had to be sawed in winters and had with boiling tea or milk. No other way. Oranges, even in summers, froze and became anti-personnel grenades. You could kill a man with them. The only way to eat them was to boil them and then they were not edible anymore. Beer would freeze rock solid too and had to be melted to drink, no good thereafter anyway. Troops would often joke that on the glacier if you 'piddled' outside in winters, the urine would freeze in the air before it hit the ground, turning into an icicle from the dick to the ground.

In our makeshift tent, a parachute curtain served as the partition between the living room and the toilet. It was not sound proof and only provided psychological privacy. You could hear the officer breathing inside the toilet and much more. With a lantern inside, the body silhouette was clearly visible with all the finer details. The bottomless crevasse served as the make shift toilets. Skid board with a hole, secured over the crevasse with angle iron pickets was the commode.

Officer answering the nature's call had to sit holding the strap secured to an iron picket, just in case the crevasse opens up or the skid board skids into the dark bottomless pit.

Second Lieutenant Jimmy, a very fair officer was a good soldier with a great attitude in life. He had just returned from a long range patrol on a glacier considered impassable. The long outing had given him terrible sun burns. Skin from his nose and cheeks was peeling off. He looked rather dark and quite unrecognizable.

Another youngster, Second Lieutenant Bishnoi had come down from Bilafond La Company after a long time. He was a darkish fellow. The terrible sun burns had made him even darker. Bilafond La was a tough assignment where everything had to be carried man pack for hours from the dropping zone. Kerosene oil was barely enough for cooking, let alone bathing, for which water had to be got by melting ice. Bathing was too much of a luxury that could not be afforded.

Arrival of own officers at Kumar was always a cause for celebration. Captain Tarun Kaul ordered beer. A troupe of three soldiers trooped in with a crate of frozen beer bottles and a bucket of boiling hot water, a standard drill for beer parties on the glacier. Everyone put his bottle in the bucket and enjoyed the flat beer as it melted slowly. The bash up was followed by a delicious lunch of rice, tinned keema and canned fish.

All done, Tarun said, "You buggers are stinking, not bathed for months. Your hot water is ready."

Bishnoi said, "I have not had a bath for over eight weeks. I will go in first and will be back soon," and disappeared into the toilet where the sahayak had set up hot water for him.

As he poured a mug of hot water over himself, he yelled, *"Oh ben--, yeh toh bahut garam hai yaar, main toh jal gaya."* (Oh shit, the water is bloody hot, I have got burnt). As he proceeded carefully, the dark Bishnoi started singing the film song *"Pani mein jale mera gora badan."* (My fair body is burning in hot water).

Friend Jimmy, never short of ideas, butted in timely with another film song *"Abe kaloo- dheere dheere bol koi sun na le..."* (You blacky - sing softly or someone will hear).

So life can be fun no matter where one is. It is all about the right attitude and it is all in the mind.

Saving the Snow Vehicles

In the vast expanse of the glacier, the snow vehicles were the life line. Unfortunately, in those early days, they mostly remained out of order for lack of spares and expertise, being new equipment. The snow scooter moved a bit, but only in reverse gear. All this inflicted extra workload on the soldiers. The dropping zone (DZ) had to be cleared manually. In those altitudes, one dropping sortie by two AN 32 transport or one AN 12 aircraft meant the company was busy for a fortnight clearing the DZ, as the loads fell over a large area full of crevasses. All this had to be carried man pack to company posts which were many hours trudge in crevasse ridden glacier over altitudes touching 19-20000 feet.

The kerosene oil (K oil) was the most important commodity,

as food was cooked on K oil burners, and it fuelled bukharis and stoves that kept us warm. Water was obtained by melting ice, a tedious time and K oil consuming proposition. Bathing in the forward companies was a rare privilege, as K oil was barely enough to manage the cooking, being humped over tens of kilometres. Men bathed no more than once or twice in their tenure of six months; they made do with wet towels at best.

Troops on the glacier were entirely air maintained. While air supply drops by Air Force transport planes brought hope and cheers, it was a risky affair. A wrong judgment by the pilot or strong winds drifted the parachutes dangerously close to the living area of troops. Besides, clearing the DZ on man pack basis was a real painful task at that altitude as all the snow vehicles had broken down. The only snow vehicle that worked was a snow scooter which only moved in reverse at a snail's pace, quite useless at best. A faulty drop by the aircraft meant an additional fifteen days effort for the men.

One day the aircrafts dropped the load way off the DZ. One load fell on a snow vehicle parked way off, smashing the driver's cabin and narrowly missing the driver. Many loaded parachutes sailed over the living area sending the troops scurrying for their lives. Since you don't have anything worthwhile to discuss at that place, you discuss any event that occurs. I narrated the incident to Major Sushil Negi, the company commander at the Bilafond La post, located in one remote part of the glacier. His advice after years of experience:

"Our Air Force is the biggest threat to own troops on the glacier. The only way you can save the snow vehicles from them is by using them to mark the DZ, right in the centre. That is where they will be the safest."

I tend to agree with him. Ever seen a four-ball on the golf course. All four aiming at the centre of the fairway and none getting anywhere near it, however broad the fairway.

Putteeram Deori and his woes....

Life on the Glacier was tough anyway, even for the fittest. For those who fell sick, it was sheer torture. Guardsman Putteeram Deori suffered from piles worsened by the harsh conditions on the glacier. Doing routine military chores with this problem could not have been fun. Keeping him off such strenuous chores did not help much. The morning ablution by itself was an exercise in endurance in those glaciated freezing conditions.

Suffering badly and unable to bear any more, he reported sick. He was evacuated to the base at least four times from a forward post. Helicopter effort being rare, he had walked up and down every time. That itself was a buggery even for a fit man. The walk one way was six days from his company, each day a gruelling eight-ten hours walk over crevasses and ice walls, negotiated with risky make shift bridges of aluminium ladders.

When he apparently got better, he was sent up the fifth time from the base. Our officers would interact with men a great deal. That helped us understand their problems better. Jimmy, a young enthusiastic subaltern saw him sitting in a tent, apparently depressed and dejected.

Jimmy asked him, how he was feeling. Deori narrated his long catalogue of woes, how he had been unwell for long and this was the fifth time he had walked up and down in this condition. He said he was fed up with life and wanted to commit suicide by jumping into the nearest crevasse. Jimmy, a humorous character said,

"You idiot, you have never been available to the battalion and been shamming all along. And now, even for committing suicide, you are looking for the nearest crevasse. At least now walk some distance to die, so we don't waste time looking for you."

That was the lighter side of course, the soldier was looked after and realizing his long lasting health problems, he was sent to the unit base on a sedentary duty.

Someone is Worse off ... Some where

On Siachen those days, apart from other problems, communications from the battalion headquarters to the company headquarters was problematic. You were connected on line, maintaining which was a major headache. The avalanches and the crevasses disrupted the lines frequently, making a difficult task almost impossible. The cable was rarely recovered and fresh cable had to be laid every time. After a while there was shortage of cable and one would invariably rely on radio. The problem with radio was, it was all in clear, and Pakistani posts heard all we spoke. Theoretically we were required to change the frequency every week. Practically that was not possible. Half the posts were either many days walk or were air maintained. The signal instructions never reached half of those posts, rendering them out of communication. The result was that the radio frequencies were never changed. As much as the Pakis eavesdropped on our communications, we also monitored all their conversations, and both sides knew that, and could do precious little about it.

Those days some officers used to be posted there for a three months tenure and were called HRM (High Risk Mission) officers. They beefed up the unit officers strength. There was one paratrooper officer belonging to the brigade commander's regiment. He was lifted by the commander in a chopper and dropped on a remote air maintained post. He was to stay there for a week after which the commander would personally take him out. The commander forgot about him. The week turned into weeks, weeks into months and there he was perched up there for more than three months. Quite frustrated, he would get on the radio and crib about his undue long stay and how he had been planning to go on leave and had been stuck there.

One day when he was on a cribbing session, a voice interrupted his conversation. The voice said, "I have been listening to your story since long. You are lucky friend. I was also sent here for seven days and I have been here for over five months now. They tell me my relief is being selected at Islamabad and he would be here in six weeks after acclimatization. That makes it seven months. But I don't trust them. I am prepared to retire from here."

"And who are you, may I ask?" asked the paratrooper.

"Captain Akhtar, Pakistan Army, on the post opposite you," replied the Paki officer. The Indian Captain felt much better. There was someone worse off.

Nunnu's Dilemma

Some real life incidents in the army can be funny and serious at the same time. Ladakh is a rugged high altitude region in northern part of India. It is characterized by lack of vegetation and harsh terrain. The average altitude of the region is over fourteen thousand feet with peaks kissing the sky. The people are very simple and hardy. Bulk of the population lives in altitudes above twelve thousand feet.

The tough and hardy soldiers from this region called 'Nunnus', meaning younger brother in the local lingo, mostly serve in a Regiment called Ladakh Scouts. Nunnus are known for extreme physical fitness and high standards of military discipline. They are born and spend their lives in high altitude region.

One nunnu while serving on Siachen glacier fell sick and had to be evacuated to a hospital in Chandigarh. After spending almost eight weeks in the military hospital, he was downgraded to low medical category and granted four weeks sick leave.

The doctor told him that because of his medical category he cannot go to high altitude areas. The simple nunnu stood confused and bewildered. The doctor asked him what the matter was. Nunnu replied,

"I was born at fourteen thousand feet. I have spent all my life at that altitude. You have given me sick leave and tell me that I cannot go to high altitude. My home, family and children are all in the village at fourteen thousand feet and if I do not go there, where do I go?"

The doctor had never faced such a dilemma before. The nunnu had no choice but to go back to high altitude, never mind the medical advice. Needless to mention nothing went wrong with him in his homeland, how could it?

Balls all of a Special Mission

Major Awasty of 5 GUARDS, a perfect officer and a gentleman, was commanding a company in Kargil in J&K in mid sixties. The situation on the line of control was tense. The company Senior Junior Commissioned Officer (JCO), aspiring for honorary rank was trying hard to impress his company commander. One day the JCO offered to organize a raid across the border to get some weapons with the help of local goondas who were past masters in the art of crookery.

Major Awasty gave his consent but the JCO himself would have to lead the raid party. That bolt from the blue was the last thing the JCO had expected. He tried to wriggle out citing multiple excuses, but to no avail. Caught in his own trap, he brought his raid party -the local thugs - for introduction with the company commander and demonstration of their martial skills.

The first was introduced as the local dada. Not a leaf moves in the area without his consent, a sickly looking five foot f-all of a chap. The second was an expert chhuri baaj (expert at wielding a knife). On a subtle indication from the JCO, he took out an ancient two inch knife about to fall apart and swished it through the air like a sword with the expressions of a samurai. The last of the special mission gang was a famous lathayte (expert at wielding a bamboo pole as a weapon). On a signal from the JCO, he showed his antics with the bamboo pole, staggering and falling in the process.

Major Awasty was amused at the entertainment show. Not satisfied with the combat potential of the 'Special Forces Team' led by an incompetent JCO, he let them go anyway. The D day finally arrived, and after much briefing and rehearsals, the raid party proceeded on its special mission. The three musketeers from the village linked up on a bridge as per plan. The party crossed the border, but paralysed by fear, refused to go anywhere near the enemy post. The JCO was worried as to what he would tell the company commander whom he had promised the sky. It was decided that something had to be taken back to show his boss, no matter what it was.

In the morning, Major Awasty stood at the observation post waiting anxiously for the high profile raid party. He saw in the distance a group of four people trudging up the hill lugging a cot, with a cow in tow. He wondered who the man on the cot was and what the cow was all about. Worried, he waited anxiously. The party reached the post huffing and puffing. As they lowered the cot, the biggest crook fell off the cot and crash landed on the ground. He lay there in pain, clutching his b**ls. Major Awasty was told that this cow was the one that the soldiers from across the border were trying to steal. Though they managed to escape, this cow was retrieved from them after a hot chase. The hero of the raid who had done a good job had fallen during the chase and hurt himself, and that is why he was brought on the cot.

It was not much later that Major Awasty learnt the truth. What had actually happened was that when the party backed off the mission, they fabricated a story around a cow. They entered the house of an old couple to steal their cow. The animals in these areas are tied inside the cow sheds or in verandahs due to extreme cold. To prevent the couple from shouting, the thugs decided to gag them and entered the house with their faces masked. The big dada grabbed the old couple by their throats. As the second one went out to steal the cow, the old courageous lady, sensing her chance, grabbed the b**ls of the dada. The crook shrieked jumping up and down but she wouldn't let go. She grabbed a sickle and threatened to chop them off as he froze paralyzed by fear. After much pleading by the team which had reached the house, she let his nuts off. By then the big dada was nearly castrated and unable to walk. They literally caught the old woman's feet and borrowed their old cow and a cot for some money for the night. So that was the historic end of the special mission.

The JCO could not look the company commander in his eyes and never suggested anything again. He knew his hopes of honorary rank had disappeared the moment the lady had squeezed his raid party's nuts. Major Awasty was furious at the harassment of the old couple. The unit had unclaimed cows in the cattle pond of the unit. He sent them back the next day to return the old cow with three more and some money; a windfall for the old couple.

Much later, I had the privilege of serving as the ADC to General Awasty when he was a division commander. A wonderfull couple, they always introduced me to their acquaintances as their colleague and not as ADC.

Indo- Pak Cooperation at Company Level

Indian and Pakistani Army have a peculiar relationship that is beyond description. The two have fought many wars. Pakistan continues to assist terrorist infiltration into India and is fuelling the proxy war in J&K. They face each other in an eye ball to eye ball scenario and cross border firing is common place. Yet, on the Line of Control, they exchange casual chat, sweets, a cup of tea and rum at times (something Pakistan Army would never acknowledge).

Major Dalbir from the Gurkha Rifles was commanding a company on the Line of Control. The Pakistani JCO on the other side called aloud for Major Dalbir.

"Saab ek request hai. Kal tussi subah dus baje meri post par jabardast fire karna saab." (Sir, please fire heavily on my post tomorrow morning).

"Kyon saab, yeh toh ajeeb see request kar rahe ho," quipped Dalbir.(This is a real strange request, asking me to fire at your post).

"Saab, gal eh see, kee, kal sadaa arms inspection ke lie AIA saab arahe seege. Mein chahta hoon kee woh na aye." (The thing is that arms inspection team is coming for inspection to my company tomorrow. I do not want them to come).

Dalbir obliged and the weapon inspection of the Pakistani company did not take place. An amazing story of cooperation between two adversaries who have fought four wars.

Backside on Fire

Our battalion was deployed in what was then East Pakistan during the Indo- Pak War of 1971. During the rapid advance, the battalion came across a brewery, abandoned in panic as the Pakistani forces withdrew.

The battalion took position nearby as the patrols were launched to locate and establish contact with the enemy. One reconnaissance party soon located the treasure.

In that operational environment, nature's call was a trip behind the bush, a bottle in hand. The call was answered in dark well before sunrise, as there was no guarantee you wouldn't find yourself pants down next to a soldier behind a bush, a bit awkward I guess. Captain Negi was a young die hard subaltern. The next morning he proceeded for a crap, as always carrying a bottle of water his batman handed him over. All done, he set about cleaning up. Before he was through, his backside was on fire. Through the darkness he tried to smell and figure out the bottle. He was in for a surprise. The batman had handed him a bottle of gin picked up from the brewery. Not many people have had that luxury before or ever since, I guess.

Another officer had a similar experience, but he was not as fortunate. He was handed over a kerosene oil bottle, and that was terrible. The itching and the smell of kerosene never left him till the capture of Dacca. Both the youngsters were conspicuous by their angled gait for the rest of the day, as they led the unit's advance towards the objective. And guess what? The battalion earned the 'Theatre Honour' East Pakistan as it accomplished its mission well ahead of schedule, thanks to these young officers, who led the advance at a rather hectic pace propelled by gin and kerosene.

Behind Every Bush is an Ambush

I was a sector commander in Kashmir. Someone asked me what was the most difficult thing that I found there as the sector commander. I said, "The most difficult thing I found was to go for a piddle when I was out on the move." He was amused but could not understand why going for a leak was so difficult.

Given that insurgency environment, a sector commander was an important target and had his own protection of about ten soldiers. They moved with him whenever and wherever he went. He also had a buddy who was his shadow. If the commander looked left, he would look left. If he looked right, the buddy would look right. The buddy would follow and trail him step for step.

The first time I was returning from my force headquarters, almost three hours drive away, I felt like going for a leak. I asked the driver to stop. The whole protection team based on three vehicles stopped, jumped out of the vehicles and took position behind the trees. As I moved towards an apple tree looking for a secluded place to piddle, the protection team encircled me and my buddy took position right in front of me, ready to shoot. I could not have unzipped in front of him, so I turned right and walked towards another tree in the orchard. The whole protection team started side stepping - with me as the epicentre. Another soldier smartly knelt behind a tree and soon my buddy had caught up with me and stood right there. Similar thing happened on my third attempt. Well, finally I had to tell them to leave me alone to piddle in privacy and peace, notwithstanding the terrorist threat. Thereon, my single point agenda before moving out was to empty my bladder.

Terrorist Troubles

Eliminating terrorists was the prime task of the army deployed in counter terrorist operations in Kashmir. While there is so much more the units do as part of counter terrorist operations, the only quantifiable part of their performance is the number of terrorists killed by the units. This, therefore, became the barometer of success of the units and by extension the commanding officers'. As the situation improved, terrorists were difficult to come by. During the briefing of a visiting officer, the last slide showed statistics of operations that showed the number of terrorists killed.

That being important for the unit's pride and recognition, the units try very hard to have this success too against their names. Given the officers' penchant for humour, there was a joke going around as to why the terrorists go underground and are particularly hard to find in the month of August. The reason- because September is the month when the battalion commander's annual performance report is written and no terrorist wants to take a chance. On the lighter side, I guess the terrorists have no respite ever, since round the year, someone or the other's performance report is due.

It is to Indian Army's credit, that the battle hardened Afghan terrorists, who came to India in early nineties with big ideas found the going tough here. They soon realized what was good for them and chose to stay where they were born. One captured terrorist described the Indian Army as a bull dog, if it ever got your smell, it won't ever let you go.

People Friendly Operations

Pakistan sponsored insurgency in Jammu and Kashmir started in 1989. I had served in a sensitive area as a company commander in the early nineties. Being initial days and the insurgency being at its peak, the army was learning the tricks of the trade. As the situation came under control, there was great emphasis on prevention of human rights violations and good troop behaviour etc. Commanders at all levels emphasized on these aspects repeatedly. At the ground level operating became extremely difficult under these constraints. Occasionally civilians would question our right to check their identity cards and refuse to show them.

Time flew by and I was posted as sector commander in an active area in 2008-2009. Things like tough interrogations became things of the past. In fact, interrogation by the army was stopped altogether and would be carried out by the police. Custodial deaths were not acceptable and it was banned to bring any civilian to the army post for any reason.

Counter Insurgency Operations in Kashmir-
WHAM -Winning Hearts And Minds

Despite the constrained environment, the units were still expected to produce results and neutralize terrorists. One day Colonel Nandal, the Colonel General Staff from the force headquarters gave me a call to remind me of the instructions against bringing civilians to the post for any reasons. I said I am aware, but why this reminder. He said, *"Apparently there has been a case of custodial death in one unit today. They were interrogating their own source."*

Funny, I thought, "custodial death interrogating own source!" Wonder what kind of own source was he. Well it was a humour in uniform.

Sometime later, there was an incident where two civilians got injured in cross-fire in a village when the terrorists fired on an army patrol. The incident took an ugly turn and remained in the headlines for months. The army came under pressure to shift the post away from the village. The corps commander was very particular on human rights issues and troops' conduct with local populace and addressed the officers. Talking to the officers, he said there are three important things that need to be kept in mind during conduct of operations. These are *Josh* (enthusiasm and go getter attitude)- *Teji* (speed in execution)- *Sidhai* (correct direction i.e. appropriate action in a given situation, ensuring no human rights violation and collateral damage). Translated in English, sidhai means straight. He said in the incident above, there was josh, there was teji but there was no sidhai. The patrol returned fire on terrorists in a crowded area where the risks of civilian casualties are always high.

I found that to be a good line for our troops to follow. I had a board put up in every company post stating basic guidelines for operations in the changed environment. The first guideline was Josh-Teji-Sidhai. During a visit to a company post I asked the senior JCO what he understood by these words. He said, josh means the platoon must do the task with enthusiasm. Teji, he explained, meant speed in execution and then chose to remain silent.

"What does sidhai mean?" I asked him.

"Sidhai means," he said, "Jab hum ambush lagate hain ya search operations karte hain toh sab jawan ek sidhai mein hone chahie." (When we lay an ambush or do search operations all soldiers should be in a straight line).

Who is she?

The Afghan terrorists were in Kashmir in the earlier part of Pakistan sponsored terrorism in early nineties. Soon they realized that Indian Army was a different cup of tea. They were like leeches, if they saw a terrorist, there was no way he was going to get away. Soon they realized what was good for them and their numbers dwindled. Those days the terrorists were ruthless with civilians. The schools were burned down and various diktats were issued. One was on compulsory wearing of burkas by women. The terrorists would wear burkas to pose as women and escape the army cordons. Soon we learned the trick of the trade and started checking hands of burka clad women to ascertain.

During an operation, 8 JAK LI had taken the head clerk along for the commanding officer (CO) to catch up with his office work. A group of burka clad women were walking away from the cordon. The sentry checked the hands of a few women and let them go. A gust of wind blew the burkas exposing their garments. The head clerk who was waiting while the CO read a letter saw something. He dropped all his files and bolted towards the burka group. The CO shouted, "What happened saab?" The head clerk shouted back "Saab, she is not she, she is he", and returned with a burka clad terrorist under his arm pit. 8 JAK LI is a wonderful battalion of Siachen fame. Everyone there wanted to kill a terrorist.

Disciplining the Tennis Balls

CSD: Canteen Stores Department- a unit run canteen that caters for provisions for soldiers.

My sector headquarters was located in a remote area in Northern Kashmir. PT and games were regular features. We had constructed a tennis court and would often play in the evening. The other officers had started learning the game there. Soon they picked up the game well and looked forward to it every day. Colonel Sudeep, our camp commandant, bought top of the brand tennis racquets for all-Babolat, Head, Prince and Wilson. Seeing them, any one would believe we could compete in a grand slam, but I guess the similarity ended right there.

The sector CSD had Cosco tennis balls and we enjoyed playing with them. All of us being novices, we had no idea of which balls to use on a cemented court. Sudeep had thoroughly got engrossed in the game he had learnt at forty eight. During his leave he took tennis coaching lessons and brought three dozen Penn balls. These were low pressure soft balls that felt great on the cemented court. At the rate these balls were being hit for sixes, soon we ran out of them and were back to Cosco balls. It was only then we realized the difference. These balls felt like stone and gave all of us tennis elbows. Some officers went on self imposed rest and tennis court was less crowded. The CSD, somehow, had two hundred Cosco balls in store and no one was interested in them. I decided we needed to play them out or else the CSD would suffer a huge loss.

One evening, Major Kuldeep, an ever cheerful Texan, was waiting for his turn to play. I saw him seriously studying the two types of balls, pressing them alternately. He sent a jawan somewhere who returned after a while. Kuldeep then tossed a Cosco tennis ball and said,"Sir, try this, how does it feel?" We knocked around with this ball for a while and wow, it felt great, just like the Penn balls. "How did you do that?" I asked. He held up a syringe saying he had removed some air from the ball with it. How wonderful, a stupid idea that was the perfect solution to the problem that suddenly made all the rotting Cosco balls relevant. As the news spread, all our tennis elbow casualties were back on the court. Thereon, the syringe became part and parcel of our tennis kit, not for a medical reason though. Soldiers who helped us during the game became masters of syringe treatment and did a great job every time a new set of Cosco balls were brought. Life was back on the court and the Penn balls were forgotten.

One day a new set of Cosco balls were brought and were subjected to syringe treatment by the soldier. As we played with the new balls, they hardly bounced. Far too much air had been taken out with the syringe. We tossed the balls out disappointed that three new balls had been ruined.

We were back to our old bald Penn balls that had hardly any cloth left over it. A little later, the soldier tossed the same balls back and asked us to try. We knocked with those balls and the lousy bounce was gone. They bounced well and felt great. The secret was simple. The soldier had injected some air back into the balls with the same syringe. Would you believe how the soldiers can improvise?

As the monsoon came, our tennis was disrupted. The balls would get wet and heavy and broke the racquet gutting. In the heavy rains, there was really no way you could play. My deputy commander said we can play after light rains. I said, "The balls would get wet from the grass around the cemented court." He said he had made some arrangements. I wondered what that arrangement was.

As we played, the balls kept getting wet from the grass around the court. But time and again dry balls appeared from somewhere. "How many balls have you got here?" I asked.

"Four," he replied.

"How are these dry balls appearing all the time?" I enquired.

"Microwave oven," he replied coolly. A soldier on duty would disappear with wet balls, bolting off to the nearby officers' mess and back after a forty five seconds microwave treatment. This treatment, while drying the ball alright, made it even harder and bouncy and our tennis game was closer to volleyball, good enough for the place we were in and we enjoyed every bit of it, everyday.

The Simple Simons from the North-East

Indian expanse and diversity are well known. The Northeastern region of the country has very remote places where one has to trek for days to get to some villages. The men from this region make excellent soldiers. Their simple, hardy life far from the materialistic and maddening metros makes them perfect material for the armed forces. In those remote places people live a happy and honest life, content with bare necessities.

A Simple Simon from a remote village in Nagaland had joined the army. Having gone on leave for the first time, he did not return on time. Another soldier from a nearby village was sent on short leave to take stock of Simon's whereabouts and get him back. On enquiring on his reasons for returning late from leave, he said he did not realize that his leave had finished. He tried to find out the date but could not as there was not even a single calendar in the whole village. The commanding officer found a novel way of overcoming the practical problem. In Nagaland the ritual of attending church on Sundays is observed religiously. The young and old flock to church in their best attire and enthusiastically sing carols supported by stylishly turned out bands. Before proceeding on leave they were told the number of Sunday church they would attend during the leave. In far flung remote villages perched on hill tops, apparently without calendars, that was a simple and effective way that worked well. No excuses could be tenable, even non availability of calendars.

In the seventies, now and then, there were cases of soldiers from this region returning late from leave. Another Simple Simon from the Northeast returned late from two months leave. He was marched up to the company commander for disciplinary action. When asked why he was late, he replied, *"Saab, mein toh ghar se time par nikala tha. Ghar se pachass metre jakar meine ek bara galti kar diya. Meine mur ke peeche dekh liya. Aur mera memsaab ne bahut pyar se aise ishara kar diya."* (Saab, after my leave I left my home on time. But I made one big mistake. Fifty metres from my house, I momentarily looked back and my wife gestured like this with great love), sweetly demonstrating the gesture of calling him back with a cupped hand. *"Aap batao saab, isme mera kya galti hai. Aap ka mem saab aap ko aisa pyar se bulata toh aap kya karta, batao na?"* (Now tell me, what is my fault in this. If your wife had called you back with such love, what would you have done? Tell me). Great dilemma for the company commander. I guess he would have taken a sympathetic view for his truthfulness.

A Bolt From the North-East !

Indian Army is a fine example of national integration. The troops come from all over the country living in perfect harmony. We have a regiment which has troops from the Northeastern states of the country. They speak different lingo at home, are not too well versed with Hindi and speak with an accent. They have devised their own terms to explain things that they find difficult to express.

A General officer visited a forward post near Nathu La on the Chinese border where a battalion of this regiment was deployed. The post was on 'Stand to' as per the procedure for the visit of a VIP. The General walked up to a jawan manning the LMG bunker and asked him. If he sees enemy coming, what are the various ways by which he can indicate their location to his section commander. The jawan rattled out all that he was taught in training.

The General wanted him to say that a target can be indicated by firing a tracer. He goaded the soldier, saying if the section commander did not see the target despite using all these methods, what else he could do. More goading, yet another method but not the tracer method that the GOC wanted to hear.

Finally the General asked, "Can you use a tracer bullet to indicate the target?"

"Haan saab - woh bhee kar sakta hai," (Yes, I can do that also), replied the jawan.

"Ap ko malum hai tracer kya hota hai, kabhee fire kiya hai?" (Do you know what a tracer is? Have you ever fired one?)

"Haan Saab, thora- thora malum hai," (Yes, I know a bit what it is), said the jawan in his Naga accent.

"Batau tracer kya hota hai?" (Tell me what a tracer is?) asked the General.

The Naga soldier cleared his throat, thought for a while and answered with a straight face, *"Tracer woh goli hota hai srimaan, jisko fire karne se uske chuttar se aag nikalta hai."* (Tracer is that round sir, which, when fired, leaves a trail of fire from its arse hole).

He was dead right, but the General did not know where to look. The officers and men of the battalion raised no eyebrows. In the local lingo used by the troops from the North-East, "Chuttar" means rear or bottom of everything.

We have the privilege of having these boys in our battalion

as well. One day I was playing volley ball with the unit team. The young smasher was smashing the ball out of the court all the time. The senior's advice to the younger one, *"Arre yaar, ball ke chuttar mein marega toh ball bahar hee toh jaega. Uska sir mein maro, phir andar ayega.* (Buddy if you hit at the bottom of the ball, it has to go out of the court, hit on the head of the ball and see).

Court Ruling Spells Trouble for Subordinates

It was in 2009 that the issue of gay rights came up in India in a big way. In the place where I was posted, we really were amused that the issue was referred to none other than our simple, inimitable Prime Minister, Mr. Manmohan Singh. Rightly, he had passed on the troublesome baby into someone else's lap.

Well, in July 2009, the court observed that IPC Section 377 relating to same sex relationship violated certain provisions of Indian constitution. Its ruling decriminalized same sex relationship that kicked up a big debate in the country. In the army, there is a bit harmless 'unparliamentariness' in our day to day language. When a senior wants to severely admonish a junior, he would say it something like this:

I will kick your butt or I will kick your arse, or I will screw you or I will have you or words to that effect. My deputy commander warned all officers as he read a SMS that was floating around after the court ruling. "Gentlemen better watch out, with the court ruling in force, the commander now can legally take your arse."

COURSES ARE A FUN PLACE

Training is the best form of welfare, they say in the army, as it saves lives. The army trains hard in a variety of ways. The army has training schools all over the country that run courses of instructions to train officers and men in every field of their expertise. It is here, either by design or by default that one experiences some great moments of fun that remain a part of one's memory. The officers understand that good performance on courses is one of the important stepping stones to a successful career. Many interesting things happen in that process, as we shall see. Battalions have worked out procedures to train their officers and men before sending them on courses, called pre-course training. Some units maintain a good collection of course material which is useful for preparation and has come to be known as 'Pre-Course Knowledge' (PCK). The PCK by itself is a bone of contention. The schools of instruction feel that PCK prevents original thinking and the students tend to give out the so called DS solutions (recommended solutions) and do not apply themselves. That battle will go on as the schools keep on changing their exercises and the PCK finds its way back to the students in no time.

An officer starts his career with the Young Officers' (YO) course. This is the first time he is out of the strict regimentation of the unit and is the boss of all he surveys, far from the strict vigilance of his unit officers.

A collection of enthusiasts from all regiments in colourful regalia exchange their experiences of joining their units. As we started the YO Course, we had officers' cross country competition. Everyone guessed who would win the race. During the lunch before the run, Second Lieutenant Rajbir Singh, a Texan from 19 Rajput stands on a chair in the student officers' mess and challenges everyone for the run. Everyone looks up, wondering at his confidence in his ability. Then he clarifies, his challenge is not for coming first, it is for coming last. The hall goes up in cheers. Then stands another challenger on a chair at the other end of the mess hall, "I do" says Second Lieutenant Somdutt Prashar from Rajputana Rifles. The stage was set for the 'battle royale'- the battle for the last spot.

There was an unending queue of young officers as they trickled in at the finish line. Much after the whole course had finished the run and the prize distribution was about to begin, in comes Somdutt charging like a samurai with a definite sense of pride at keeping his word, beating Rajbir to the last spot. The course goes up in cheers. Then everyone is looking for Rajbir, not to be seen. And then roars in another samurai, friend Rajbir, three hundred metres behind Somdutt. He had been hiding behind a bush just short of the finish point waiting for Somdutt and his trick worked. He had won the 'battle royale' for the last spot. So, Rajbir, true to his word, was the last samurai standing. That is the kind of fun young officers looked forward to on their first outing from the unit.

Dummy Officers

We reported to the Infantry School, then in Belgaum, for the Young Officers' Course. This is the first outing of officers from the unit and the course is seen as a rest, recoup and refit opportunity. More often than not, discipline here is the first casualty. The first leg on the course those days used to be the platoon weapons leg. The student officers were divided into squads of ten and placed under the tutelage of a weapon instructor, a sergeant. The sergeants took charge of their wards and walked them to their squad post. These NCOs would be our friend, philosopher and guide for next eight weeks of the course. These were outstanding NCOs, handpicked to train officers. Some of them had great flair and good sense of humour.

Havaldar Ban Bahadur took charge of our squad and marched us to our squad post under a keekar tree, that promised no shade in the simmering heat.

"Squad, savdhan - vishram- savdhan. Dahine se ginti kar. Ek se panch tak Number 1 Group, baki bacha hua Number 2 Group. Number 1 Group ka Number 1 squad senior hoga. Kisi shriman ko koi shak?"

(Squad, attention- at ease- attention. Count from the left. One to five –Number 1 Group, remaining are Number 2 Group. Number 1 of Number 1 Group will be the squad senior. Any doubts?)

"Koi shak nahi toh class soche aur bataen mera naam kya hai?" (If no doubts, then someone tell me what is my name?) and smartly covered his name plate.

None had noticed that since it was the first time they had seen him. So he proceeded, *"Saab thora chatak ho jao. Aisa dheela officers sahiban fauz ke lie theek nahin hai."* (Sir, wake up, such inattentive officers are not good for the army). He then asked all officers to introduce themselves by turn starting from one end. Everyone did, giving out their names and units. In the middle was one officer who introduced himself as Second Lieutenant DP Singh. The instructor butted in, *"Kya naam bataya saab?"* (What did you say is your name, sir?) *"Second Lieutenant DP,"* replied the officer.

"Low saab, kal tak toh DP Rifle aur DP LMG hee suna tha, aaj kal afsar bhi DP hone lage. Kya hoga is fauj ka." (Look at that, so far we only had DP-rifles and DP- LMG. Now we have DP officers. God save the Indian Army). DP stands for Drill Practice. DP rifles are dummy rifles used only for the purpose of training and cannot fire.

The Safest Place on the Range

Young Officers' course was performing so badly in firing that one instructor commented; "When your course is firing, the safest place on the range is behind the targets."

How Many Men in a Three Men Trench?

A question- answer session in an officers' weapon training class:

"Sochen aur batayen ki three men fire trench mein kitne jawan aate hain?" (How many soldiers come in a three men fire trench?)

"Ek saab." (One)

"Galat, is ko durust karenge, koi aur cadet?"

(Wrong, anyone else?) The instructor would ask.

"Doh saab." (Two)

"Takriban theek, isi ko aur clear karenge bayen se number teen."

(Generally correct. This will be further corrected by the officer third from the left). The instructor would persist.

A funny officer would reply, *" Mote doh aur patle teen."* (Fat two, slim three).

The instructor would not know what to do next and would wind up by saying, "Remember sir, three men are authorized in a three men trench."

Bash on Regardless

During the Platoon Weapons course at the Infantry School, we were going through the instructional practice test. Here a student is required to take a weapon training class. Those performing well in this test are graded instructor. The student being tested was taking a class on handling a 2" Mortar- loading and unloading and firing the bomb.

While giving the demonstration on loading and unloading a bomb, he loaded a bomb into the barrel of the Mortar and continued his monologue. In the process he forgot to remove the bomb from the barrel and never realized it.

Later, during the class he was to give a demonstration on unloading a misfired bomb (bomb that does not fire due to some defect). To demonstrate the method of unloading, he loaded a bomb and showed the procedure of unloading it.

"Yaddi misfire bomb ko khali karna pare, ya bina fire kiye khali karne ki jarurat pare, toh dahina hath barrel ke upar rakhte hue bomb ko is prakar khali karen aur ek taraf rakh de." Then, as he straightened the Mortar, he heard a small thud of a bomb inside the Mortar. Realizing that there was already another bomb in the Mortar tube, he continued unruffled.

"Is prakar, yadi mortar se ek aur bomb nikale toh use bhi sikhe hue tarike se khali karte hue ek taraf rakh den." (If you are required to unload a misfired bomb, put your right hand on the muzzle, turn the Mortar upside down and remove the bomb. If you find that there are two bombs inside the tube, unload the second one also as taught and keep the bomb aside). He was awarded an instructor grading.

Commando Course

The commando course, now called ghatak course, is a compulsory course for infantry officers. It is not so for other arms and services. It challenges an individual's mental and physical endurance limits. It makes you realize that your endurance limits were well beyond what you thought it was. That is a great learning. Physical endurance is one important part of the course manifested by speed marches etc. You have unarmed combat classes of eighty minutes, in which seventy minutes are devoted to high voltage warming up. With every passing day, the physical strength keeps depleting and by the time the course ends, the commandos are surviving on pure will power. The instructors are young in service and their enthusiasm does not help the cause of the commandos either. All this notwithstanding, the commandos have fun and enjoy the course as you shall see.

The Compass and the Pull Through

During the commando course, we had a Senior Instructor (SI) from Sikh regiment. He was a foul mouthed character. Let us call him Major Sandhu. The young officers from the Sikh regiment would joke, that in their units, use of flowery language was a regimental tradition. And, unless you can abuse well, you are not a good officer. Major Sandhu had put in almost fifteen years of service, as such, foul language was his second nature, flowing effortlessly like Beethoven music.

During one outdoor exercise, the SI noticed the commandos carrying military compass (instrument used for navigation) in their pockets without securing with a string. This was contrary to orders as they could get lost. The commando course provided perfect environment for losing anything unsecured as the commandos were always on the run, day in-day out.

He directed that next day he would like to see all compasses secured properly with a 'sutali' (a cotton string), which he pronounced as 'sutadi' in typically Punjabi accent. A standard sutali is actually made of jute fibre.

The next day we brought the compasses secured with sutali. Major Sandhu ordered us to hold our compasses over our heads so that he could see them.

We stretched our hands up with compasses secured with a jute sutali. Seeing the jute strings, he blew his top.

"That is not what I wanted you idiots. Buggers, when I tell you to bring a sutadi, I want a nylon sutadi and not a jute sutadi. Got it? Make no mistakes about it or you'll forget the day you were born. I will see your compasses tomorrow again. Anyone who has a nylon sutadi?" he enquired.

Now, there was an officer, Commando Kumar who had failed in the commando course twice before. He was a short service officer due for screening for a permanent commission. He needed to pass the course at any cost or else he would be thrown out of the army. He was trying hard despite his physical limitations, attempting to please the instructor whichever way possible. He yelled "Yes sir, I have a nylon sutadi." Kumar sensed the golden opportunity to make a mark, or so he thought.

"Good, Kumar let me see your nylon sutadi," said the instructor. Before the SI could finish, Kumar was already on the move and sensing a kill, eagerly stepped forward and brought out his compass. The compass was secured with a rifle 'pull through'. ('Pull through' is a nylon cord with a metal weight at one end, used to clean the barrel of a rifle). The SI stared at the pull through for a while in silence and then started with his stone dead expression.

"OK, so this is your nylon sutadi. Good."

"Yes sir," replied Kumar, visibly pleased with himself.

The SI then proceeded with his dressing down of Kumar, in a manner, so comical yet so humiliating, that one wondered whether to laugh or cry for Kumar.

"So this is what you have brought to show me. A rifle pull through to secure a compass. Kumar, do you know that you are the biggest bloody Charlie (actually Hindi equivalent spoken) born on this side of Suez Canal?"

Kumar, whose world had suddenly come crashing down, seemed to have gone into a shock. He had too much at stake and did not want to annoy the SI at any rate.

"Yes sir," replied Kumar.

Holding the pull through in his hands, Major Sandhu looked at Kumar. "Do you know what is this, Kumar?'

"Yes sir," replied Kumar.

"What is this Kumar?"

"Sir, a pull through."

"Do you know what it is used for Kumar?"

"Yes sir."

"What is it used for Kumar?"

"Sir for ...." and before Kumar could speak further, the SI would interrupt "for pulling through, Kumar."

The SI then held the two ends of the pull through, the upper hand in front of Kumar's face and the other at the level of his belt. And then gesturing with his hands, "You know Kumar what I will do? I will put this 'pull through'- through your upper hole, take out from you know where, and then pull it..." and he would pause and Kumar would say "through", all the time gesturing up and down with his hands to time with Kumar's "through". And this comic street play continued for the next few minutes as Kumar kept saying "through" after the SI's "pull it." It was real 'tragicomedy'. We were barely able to keep control of ourselves.

On the other hand, the pitiable state of Kumar simply cannot be put in words. Even Shakespeare would be at a loss of words to do so. The fun aside, I tried to place myself in Kumar's shoes and felt sorry for him. For a man whose entire career was at stake, an opportunity, which he thought would do him good, backfired.

I do not really know what happened to Kumar or whether he was retained in service, but have never seen him since.

We suffered Major Sandhu for the balance of the course. Many such incidents gave us a reason to laugh and raised our spirits despite the bone breaking commando schedule.

Vini, Vidi, Vinci- I Came I Saw I ...

Commando course is a much feared course, thirty days of unadulterated buggery. It is compulsory for infantry officers. By the end of the course, one realizes that the body can take a lot more than one thinks. Amongst many exciting things, if I can call them so, is the Commando jump. It involves jumping in the pool called 'Lido Tank', climbing a high iron ladder for a balance walk on a foot wide plank perched at an uncomfortable height of about sixty feet. A fall would bring you splashing down into the 'Lido Tank'. Half way through is a three step stairs that shakes when you walk over it. At that height a small shake is worse than an earthquake measuring eight on Richter scale. On that God forsaken place is written in bold Julius Caesar's quote, "Vini, Vidi, Vinci" meaning "I came, I saw, I conquered." For some commandos, it felt more like saying "I came, I saw and I ran".

After the balance walk follows a monkey crawl on a horizontal rope another ten feet higher, over sixty feet above the water tank. A commando does monkey crawl till the centre of the pool indicated by a red cloth. He then hangs there, legs stretched tight and shouts "Commando XYZ, ready to jump sir." The instructor below shouts "Commando XYZ - jump," and fires a gun shot. The commando is required to leave his hands at the gun shot and drop into the tank sixty feet below, hands stretched above the head.

Perched at that height, one can see the whole town. The commandos waiting down below look like Lilliputians. A fairly huge water tank looks the size of a handkerchief. One wonders whether one will land in the tank or crash outside on the concrete pavement.

One can see the looks of concern on commandos waiting for their turn. Those who had finished were the last word on how to negotiate the worst day of their life. "Pretty simple yaar, just leave your hands at the gun shot, gravity will do the rest."

Commando Barrali was not amused with all these expert comments. Where the shucks have I got stuck, he wondered. Soon his turn came and he quietly marched to the starting point. "Come on Barrali, get a good observation," shouted his buddy. "Good observation my foot," thought Barrali. He would have been happy to get through it even with a couple of bad observations.

Finally, the moment of reckoning had come. Commando Barrali jumped into the Lido tank. Confidently climbed the ladder and there he stood on the balance ramp sixty feet above the mother earth, and that much closer to heaven. A deep breath and he was on his way. Barring a few hiccups on the steps, he walked across without much trouble. Everyone watching from below was impressed. Barrali then got on the horizontal rope and began his monkey crawl to the centre of the rope where a red cloth was tied. Like the others before him, he brought his legs down and hung there for a while. Then he shouted, "Commando Barrali, ready to jump, sir." "Commando Barrali-jump," shouted the instructor from below, followed by a gun shot.

As everyone waited, Barrali did not jump and remained hanging. "Jump you clot", shouted everybody from below. Amidst the noises from below which he could hardly hear, Barrali gathered courage and again shouted, "Commando Barrali, ready to jump, sir." "Commando Barrali-jump," shouted the instructor again, followed by another gun shot. Everyone watched holding their breath, but Commando Barrali wouldn't budge as he hung there precariously fidgeting from fatigue.

"Jump for your life you bloody clot," an instructor from his own unit yelled.

"Come on Baralli, jump damn it," shouted his buddy commando. "It is simple yaar, just leave your hands," yelled another. Hanging for his life, Baralli neither heard anything nor did the simplest of things- leave his hands.

The drama went on for a long time. Amazing that Barrali, a bit prosperous in health could hang for so long, not easy for even the fittest of guys. Too tired to do so any longer, his one hand slipped off the rope and he dangled with one hand like a sack of potatoes.

He'll now fall off thought the commandos waiting below, and may be, pass the test.

That was not what Barrali had in mind. With great effort he half pulled himself up and managed to get his hand back on the rope. He then did what no one had imagined. Despite hanging there for so long, he lifted his legs up on the rope. Against the din of all abuses and howling hurled at him from below, he started his monkey crawl backwards, heading back to the balance ramp.

"You son of a gun, what are you up to, jump you clot," shouted his unit officer. Baralli heard or would hear nothing. He reached the balance ramp after what seemed to be an eternity, slowly walked back over the ramp and came down the ladder pacing the route all over again. Thoroughly drained and sapped, he clung to the wall of the tank breathing heavily in tragic relief.

His troubles were not over yet. He was greeted by his unit officer, who out of regimental concern had been shouting his guts out, willing him to jump. With Barrali within his reach, popping him in and out of water, he yelled with rage,

"You dumb idiot, never seen a clot like you. You are a bloody coward."

"No sir, I am not a coward," replied Barrali breathing like a horse, clearly relieved with his proximity to mother earth.

"You bloody clown, you lack determination, you are not determined," shouted the instructor blind with rage.

"No sir, I don't lack determination. I am determined, but I am determined not to jump,"said Barrali with steely determination.

Survival of the Crookest

Towards the last leg of the commando course there is a survival exercise. The commandos are required to live off the land for 48 hours. No rations are issued. There is an effort on part of some commandos to smuggle a few eatables through and that is only to be expected.

Cigarettes are forbidden for the duration of the exercise. The commandos are stripped to the undys to ensure that nothing is being smuggled. During the break in a jungle, I saw two commandos smoking. Surprised, I asked them, "How did you get this through, buddy?" "The name of the exercise is survival man, you got to learn to survive," saying so he grabbed his rifle, tipped the barrel down, and lo and behold ten cigarettes popped out of the barrel.

"Wow! That is great, full points for the ingenuity," I said.

"Well, want to eat a chocolate in the middle of the jungle?"

Not quite believing, I said, "I won't mind."

He grabbed his rifle again, removed the magazine from the rifle and pulled out a couple of chocolates from the magazine". Brilliant, he had removed the spring from the magazine leaving enough space for a few chocolates.

"Tell me, want to eat some mint toffees?" he asked. After his demonstrated performance, I did not doubt him anymore. He turned his rifle upside down and pulled out a packet of mentos from the rifle butt. The mentos was a perfect fit in the oil bottle slot.

"Do you want to eat chicken? Will get you that as well." I believed him. Do what you may; the commandos always out smart you. Surprisingly, I got the impression some commandos were having a great time with food in an exercise of living off the land.

Better not Take a Chance..

The toughest exercise in the commando course is the forty kilometres speed march. This is conducted after the survival exercise, where you live off the land for forty eight hours. In the middle of forests, nowhere near a civilization, there is little solid food that you can smuggle through with all the ingenuity at your command, never mind the brilliant use of rifle space.

After forty eight hours of living off the land where you are patrolling, ambushing, raiding and trudging through filthy rivers and nalas, you are very- very hungry and almost at the end of your mental and physical endurance. You are given a hot meal at three in the afternoon and the speed march commences at six in the evening. Two days without a meal and with the monster of forty kilometres speed march looming, there is little appetite for food, and I guess, too hungry to eat.

A commando carries twenty kilo sand bag on his back which is weighed both before and after the speed march. The distance of forty kilometres is required to be covered in six hours. Considering that the commandos have had little to eat and almost no sleep in the past two days, and with every passing day eating into your strength and endurance, it is a tough undertaking by any standards. In the old days it was mandatory to pass this test for an 'instructor grading' only. While everybody started the speed march, many, not in the run for an instructor grading or up to testing their endurance would take advantage of the rule and sham it out.

Since this exercise has a high rate of dropouts due to heat and dehydration, there is one **'STRAGGLERS' VEHICLE'** with a big board in front and rear that trails the commandos.

This speed march is the last event of the course. Thirty five days of physical and mental battering takes its toll. A commando by now is surviving purely on will power by this stage. Those who fall off or cannot continue the speed march are picked by the straggler vehicle, which also carries a Nursing Assistant.

With us on the course was quite a shammer, we will call him Sangha. He considered the speed march a waste of his intellectual talent and had no appetite for such tough things of life. Waiting at the start point before the speed march, I overheard him telling his friend Dallal from Punjab Regiment.

"Guru, speed march shuru hote hee straggler vehicle pakar lena, baad mein pata nahi jagah mile na mile."

(Partner get on the straggler vehicle the moment the speed march starts. You never know, you may not find a place later if you delay).

And believe me, it was less than half a kilometre from the start point that I happened to look at the straggler vehicle following us. I saw Sangha and Dallal in the business class seat of the 'Straggler Vehicle'-the co-driver's seat in front. They took no chances and hopped in even before the vehicle had crossed the start line.

The Streaking Paratrooper & The Army Act

The paratroopers are a class apart. They are known to do the unthinkable and that is only to be expected. No wonder the Parachute regiment motto, 'Every Man an Emperor'. They have repeatedly proved their mettle in all wars and counter insurgency operations. One paratrooper showed why they are a breed apart, this time, far from the clatter of guns.

Mhow is a small town in Madhya Pradesh. It was served only by a metre gauge train till sometime ago. The horse driven tongas are still used to ferry passengers from the railway station. It is famous because of the number of army training institutions located there. Majority of locals make a living from the army in some way or the other. There is just one small market with one main narrow street that remains overcrowded every evening. That is the only place in town where the officers attending courses or the newlywed officers, normally in plenty, can pass spare time. Even walking through that crowded street is tough. Add to that scores of young officers brandishing their new shining bikes. All in all, a buggers muddle on that street.

Well, young officers can be dangerously regimental, especially after a drink or two, and would do anything in the name of their regiments. It was during the Battalion Support Weapons Course at Infantry School, Mhow, that some officers went on a drinking binge in the mess. Someone offered a case of beer to anyone who would drive stark naked through the Mhow town. A paratrooper officer, let us call him Captain KJ, pretty glad by now, accepted the challenge. He stripped naked to birthday suit, kicked his bullet motor bike and drove off. A few others followed in a noisy cavalcade to monitor that he actually did go to the town. Mhow military station has an open air theatre which is overcrowded on weekends. KJ surprised everyone as he headed towards the theatre. The houseful theatre was rudely jolted by the noisy bullet as KJ entered from one end and rode right across the screen. He then headed for the town. While driving through the overcrowded street, quite naturally, he created more than a ripple as the crowd watched in disbelief not knowing what to do.

An officer instructor, out shopping in town, had never seen such a spectacle. He was certain that it had to be a student officer and decided to follow him in his car. This streaker saw the tail and sped into a side street vanishing into the dark alley. The instructor tried to catch up, but, in the narrow streets of Mhow, he was no match for a crazy paratrooper on a bullet.

An identification parade was organized straight away. To cut the long story short, friend KJ was finally identified and his motor cycle was confiscated. A court of inquiry was ordered and the officer was charge sheeted as under:

"The officer is hereby charged under Army Act 52 E, disobedience of local orders, in that, he on 12 Feb at 2100 hours failed to observe anti-malaria precautions after sun set."

The Battle of Khajjis- KISS

- *Khajji- A palm tree. In the army, a Sikh person, not popular with the subordinates is referred to as Khajji. A simile is drawn since the palm tree is tall with leaves at the top. A Sikh wears a turban and sports long hair.*
- *D S- Directing Staff is the Officer Instructor.*
- *Bail- Ox in Hindi.*

Old adage *"When troubles come, they do not come in singles, they come in battalions,"* is really true. Captain Kumar would not have imagined how true it was till he experienced it one day.

Indication of landmarks can be tough, especially in a terrain devoid of prominent landmarks. It is often said, good indication of land marks is half the battle won. In 1982, we were doing the Anti-tank Weapons course. Those were the days of the indomitable 106 mm Recoilless Gun (RCL) of Havaldar Abdul Hameed fame, the winner of Param Veer Chakra in 1965 Indo-Pakistan war. Given its back blast and other shortcomings, the officers did not think very high of the weapon. Whenever one said so, the example of Havaldar Abdul Hameed was quoted. Lack of confidence in a weapon could not be encouraged.

We were out on a field exercise where the officers were to be trained in deployment of these guns in defence.

Captain Kumar from Kerala was soon to realize what Abdul Hameed did in the 1965 war with the RCL gun was relatively easier than what he was to do that day.

Having been given the battle situation, we were given time to study the ground and be prepared to give orders for deployment of two RCL guns in a company defended locality. As it happens on courses, only the 'Keen Kumar' student officers do the serious work. The vast majority trust their luck, not to be picked up for giving orders.

Kumar, with two 'Charlie' gradings (considered poor performance on a course) to his credit on this course already, was a serious student. Including his Young Officers' course done earlier, he already had bagged a hat trick of 'Charlies' very early in his career. Keen to break the jinx of 'Charlies', he worked hard and volunteered everywhere. As the luck would have it, he was nominated to give the orders. Delighted, Kumar jumped and stood in front of the squad raring to go.

"Section look in front," started Kumar.

"Stop," butted the D S. "Orders in Hindi, you will be briefing jawans". It was difficult to distinguish Kumar's English from his Malayalam; his Hindi was simply blue murder. Nevertheless, he continued in his typical tambi accent.

"Section samne dek. Dur doh-soh, skyline par ek akela khajji, ***naam Akela Khajji.*** *Akela Khajji our dur tak ki line, am rukh.* (Section look in front. 200 yards on the sky line, lone khajji {Palm tree}. **Call it Lone Khajji.** Lone Khajji and the line beyond is the general line of direction).

Akela Khajji daine dow baje, aath degree, teen soh, dow jurvan khajji, ***naam Jurvan Khajji.*** (Reference Lone Khajji, right two o' clock, 8 degrees, 300 yards twin khajji. **Call it Twin Khajji)**.

Jurvan Khajji daine chaar baje, unees degree, ek kharra khajji, ***naam Khara Khajji.*** (Reference Twin Khajji, right four o' clock,19 degrees, one standing khajji. **Call it Standing Khajji)**.

Khara Khajji, baine saat baje unees degree, ***ek para khajji naam Para Khajji.*** (Reference Standing Khajji left seven o' clock,19 degrees, one lying khajji. **Call it Lying Khajji)**.

Para Khajji, teen chauthai baine, barah degree, chotta khajji, ***naam Chotta Khajji.*** (Reference Lying Khajji, three quarters left,19 degrees, one short khajji. **Call it Short Khajji)**.

Chotta Khajji, dahine teen baje, aath degree daina jhuka khajji, ***naam Dahine Jhuka Khajji.*** (Reference Short Khajji, right three o' clock, 8 degrees, right leaning khajji. **Call it Right Leaning Khajji)**.

Dahine Jhuka Khajji, dahina teen baje, aath degree, ek dood dene wala bail, ***naam Dood Dene Wala Bail*** (meaning cow). (Reference Right Leaning Khajji, right three o' clock, 8 degrees, milk giving Ox. **Call it Milk Giving Ox)**.

Dood Dene Wala Bail, baine nau baje barah degree ek terha khajji, ***naam Terha Khajji.*** (Reference Milk giving Ox, left nine o' clock, 12 degrees, slanting khajji. **Call it Slanting Khajji)**.

Zamini nishaan ek baar phir se sunen. (I shall give out the land marks again).

He went on to repeat the landmarks which had no relation to what he had indicated earlier. There was complete hodgepodge of khajjis; confusion - completely - confounded. He then continued,

Mere Section ka jimmewari ka ilaka Khara Khajji se lekar Para Khajji tak."

(The area of responsibility of my section is from Standing Khajji to Lying Khajji).

And then followed one khajji after the other. By then Kumar's confused mind was a maze of khajjis. Looking to describe the limit of opening fire with the dood dene wala bail (cow), he found the cow had got up and was merrily ambling away. Kumar was in utter shambles, as he could see yet another 'Charlie' beginning to take shape. Like a good soldier, he pressed on and completed his briefing.

"Kisi ko koi shak, agar shak nahin tow ghari milaen." (Anyone any questions, if no questions, synchronize your watches).

"Class savdhan baith, I have finished sir," reported Kumar with a sharp salute attempting to put on a brave front.

To be fair to Kumar, there indeed were very few landmarks in that barren landscape other than khajjis. It was a difficult task indeed. But then, expecting the 'bail' to stay till the cows come home was expecting too much. As the luck would have it, the foundation of yet another 'Charlie' was cast in concrete that day. He completed his hat trick of 'Charlies' in that course.

Well, the DS, himself a khajji, listened patiently and then walked up front. He asked how many officers had understood the orders. No one raised his hand. They were all confused in the maze of khajjis. All they remembered was *'dood dene wala bail'* (cow), which was not there anymore, and synchronization of watches. The DS said if this is the state of the officers, how do you expect soldiers to execute the task. His advice- KISS. 'Keep It Simple Stupid'.

Short Sighted Surdy and his Spectacles

I was doing the 'Regimental Signals Officers Course' in the Military College of Telecommunications Engineering (MCTE), Mhow, with two course mates. One-Ravinder Singh, a bespectacled hairy surd known as 'Cheeku Khalsa' (chick surd), the other, Piyush Shandilya, another wonderful bloke. Shandilya had given Ravinder the name 'Solenoid', an appropriate name for one attending communications course. Three of us were back benchers of the course. One day Solenoid forgot to bring his spectacles to the class. The instructor asked him to read the slide on the screen. Taken by surprise, he got up in a hurry and said,

"Sorry sir, I cannot read the slide."

"Why not?" asked the instructor, a bit surprised.

"I have forgotten my testicles in the room," replied Solenoid.

"OK, please sit down, do not forget them again," said the officer who also used spectacles. I guess he missed the slip of the tongue.

On the first day of this course we had assembled early in the main hall. A tall handsome officer from Mahar regiment stood in front of the 'Best Students' board and announced seriously. "See that last name on the board. He is my regimental officer. He was the best student on the last course and my aim is to repeat his performance and take the second trophy to the battalion."

Very confident, we all thought. In the first written test held after four weeks, he failed. Thereafter he was always found in the students bar and stopped his loud thinking on his course ambitions.

MCTE used to hold inter-wing sports competitions. I was playing basketball for the All Arms Wing. We would practice everyday and that took a lot of our time from the course curriculum. We were given an individual assignment to make a training programme. It is a very tedious thing and takes a whole lot of time. With the daily sports commitment, there was no time left to do this. I requested my trio of Shandilya and Solenoid. They did all the hard work and made an impressive training programme. Back late from the game, I sat with their assignments and made mine.

After a few days we received our files back. Both Shandilya and Solenoid had got remarks "Good work". As for me, the instructor had first written "Excellent work", scored it out and written "Very good work". I wondered why this change of heart by the instructor.

I got a message that the Senior Instructor would like to see me. I reported to him and he asked me,

"Sengar, what kind of assignment was the making of the training programme?"

"Sir, it was an assessed work and an individual assignment," I said.

"Then why did you allow Shandilya and Ravinder Singh to copy from you? I did not expect that from you."

I did not know what to tell him. It was my team's effort and not them but I who had used their work. Well, I guess I had created a good impression on the course and I could do no wrong. Nevertheless, I told him they had not copied from me, but we had worked together to save time. My work remained "Very good" as theirs' remained "Good work."

We had a great laugh nevertheless.

The Junior Command Course

Junior Command Course (JC) is an important career course. It trains officers in commanding a company in peace and war. A good grading, that is an Alpha or a Bravo (Instructor) on the course opens avenues for good staff and instructional postings that facilitate a good career. Officers burn midnight oil and slog their backside off. For many sincere and keen blokes, it is a do or die effort and they would go to great lengths to get a good grading.

An all arms course, and a mandatory one, it has officers from all arms and services. Of all the courses, the PCK menace is the greatest here. Officers have a vast collection of old course material and 'pinks'. ('Pinks' are recommended solutions that are guidelines for the instructors to conduct the class. The students are not supposed to have these).

It is here, with officers in pursuit of a good grading that one gets to witness 'tragicomedy'. The grading system is very strict and only a small percentage of the course really gets either an Alpha or a B (I). The gradings are based on written test marks and class performance. For an Alpha grading, a student officer has to score at least seventy percent marks in both, not easy. At the end of the day, JC Course leaves many shattered dreams and broken hearts.

PCK, Boon or Evil

My battalion was posted on Siachen Glacier in 1986. Whilst being there, I was detailed to attend J C Course Serial 49, scheduled to commence immediately on completion of our tenure on the glacier. Siachen those days was a tough place, as the living conditions were at best poor. Things were only starting to happen. The freezing cold and the weather was the worst enemy. The feet were on solid ice and not on mother earth. To make matters worse, we were living in a makeshift tent made from cotton parachutes. Add to that the kerosene lanterns by which we survived. That was an environment where even reading a pondy was asking for too much, leave aside serious stuff like attack across a ditch cum bundh and heaps of training manuals.

I was not prepared for the all important career course. The unit on my request took up a case to cancel the course. Pat came the reply "Negative, officer to proceed on course, as detailed." I was not surprised.

I hunted around for training manuals and pamphlets. "No pamphlets sir. Everything is at Srinagar," replied the head clerk. In any case, my battalion never had a record of keeping past course materials. None were to be found; no sensible unit carries unnecessary weight to the glacier. On de-induction we were stuck at Leh, waiting for aircraft lift to our permanent location at Srinagar.

So, unprepared, I was on my way for JC course straight from Leh. My baggage- a rucksack with a sleeping bag, a twenty one inch leather suitcase (VIP blow-plast suitcases were not invented then), a mountaineering mattress for a bed and one bed sheet. My reading material - one pamphlet on patrolling used in Indian Military Academy.

A long journey through better part of the country took me to Mhow railway station. There I met many hopefuls who had landed for the same course. I met Captain Padam Kumar with whom I had done a course earlier. He stood surrounded by two big steel trunks, a large suit case and an oversized hand bag. A huge 'Hold all' threatening to burst at the seams, secured in the last hole of the strap – reinforced by signal cable WD, completed his wardrobe. Close to him stood his burly batman with some plastic bags containing military boots, a table lamp and a bump pad.

"Hi," he said, "long time no news yaar."

"Hi, great to see you Padam, come for the JC Course?" I enquired.

"Yes, and are you going on leave?"

"No, what makes you think so?"

"Your baggage, just a small suit case and a rucksack, so I thought you are going on leave."

"No Sir, I have come for J C course too," I said.

"Good, so your batman must be bringing rest of the baggage," he asked confidently.

"No, this is all I have," I said.

"You mean, JC course, three months long, with just this, I can't believe it. Where are your notes and pamphlets?"

"One pamphlet on patrolling is all I have, that's it," I said.

"How will you manage yaar, this is a career course. You need all the pamphlets and notes and no one has the time to share with others."

"I know, could not help it. Are there other officers you are waiting for, all this baggage?" I asked.

"No, this is all mine yaar," spelling out kit inventory of each box in detail.

"This box contains pamphlets, this one has confidential pamphlets, this one has JC course notes of my unit officers and PCK (Previous Course material) of JC course 46, 47 and 48 and some notes of senior command course, and that is my bedding. Did not take a chance, brought my own table lamp.

Important course you know, got to do well," he said seriously.

"And I notice you have brought your own bump pad too. Do not trust the College of Combat to take care of your bumps?" I asked. Inadvertently my mouth shot off, "You got too much baggage, ever heard of the saying, *'Nalayak ka bistara bhari'.*" (A dumb student has a heavy bag).

I shouldn't have said it, I realized, but the bolt was shot and I chose to leave it at that. Obviously he did not like it, I could make out.

Padam hired a coolie and moved on. The porter had to get a cart for his baggage. He offered help for mine. I thanked him, picked up my stuff myself and moved along with him to the military bus.

During the course we would meet often and have a casual chat. When the course finished and the gradings were announced, he came up to me and said, "Congratulations, you've done very well."

"Thanks, how about you?" I asked.

He said, "You remember before the course began we met at the railway station. I asked you about your baggage. And you saw my baggage and said, *'Nalayak ka bistar bhari',* I did not like it then. But now I realized how right you were. Now that the course is over, and I have got a bloody 'Charlie' again, I realize that ninety percent of material I am carrying is total shit. Why the shucks am I roaming around with Senior Command and Higher Command Course material when I am doing piddly bloody JC course? I am carrying the PCK of the last three JC courses. Believe me; I wasted hours every day trying to dig out the right notes from the big pile which got increasingly disorganized with every passing day. Half the time I never found the right one for the next day. And when I did find it, I didn't understand a damn shit of what crap they had written. Either I could not decipher their writing, or their private abbreviations, or their codes. It contained nothing worthwhile and I couldn't make out head and tail of it. And I've got a damn bloody Charlie," he said rather dejected.

"Don't worry mate, you worked hard and must have learnt a lot on the course," I tried to console him.

"Learnt a lot; my bloody foot. I cannot go back and tell my commanding officer, sir, I have come back from JC. I have got a 'Charlie' but I have learnt a lot. He will kick my butt for saying this. I am simply 'Charlie' on JC, the way I am on BSW and YO course, period. The only positive I can see is, unlike others, I have maintained my performance," he concluded on a lighter note.

I clearly saw his point.

"And what are you going to do with your notes?" I asked.

"Yah, good question. In addition to all the shit I brought, I had added some more of my own shit from this course. I have done great service to my unit officer coming next for the course. I have burnt all that shit," said Padam Kumar emphatically.

A Bridge too far

On the all important JC course, Pradhan was another officer, who, like many others, was very keen on doing well. We had known each other in the National Defence Academy and were placed in the same division on the course.

When the course started, he told me that he was determined to get an Alpha grading. Getting an Alpha on this course is never easy, as only a small percentage of officers manage to get one. I asked him, "What makes you so sure of getting an Alpha?" He replied, "My battalion officer had attended the previous course and gone with an Alpha and I have got his PCK material (Previous Course Knowledge). Also, my first company commander, now a colonel, is instructor here in JC course itself. There is no way I can be denied an Alpha." He said with supreme confidence bordering on arrogance. I agreed, the odds could not be better, but I thought he still had to get the marks, PCK or known instructor notwithstanding. I wished him good luck.

The officer worked very hard burning midnight oil. Every time I went to his room, there he was, glued to the study table attacking the pamphlets under the yellow lights of the table lamp. Before we realized, the mid-term written tests were held. Good performance in this test was the key to getting a good grading. The results of the test were still not declared. I visited Pradhan in search of PCK for a very time consuming and painful exercise relating to calculations of mule loads in high altitude operations. I found him lying face down on the bed in his ill fitting frenchies. From his roommate I learnt that he had apparently got to know his written test marks from his unit officer and was quite dejected. He had walked into the room, stripped and without saying much had slumped on the bed.

I called up his name a few times. He did not respond. I then shook him and said, *"Abe uth guru."* (Get up mate).

He turned around on his back and said, *"Abe kya uth, bade aye they Alpha lene. Bada PCK le ke chale they. Bada ghamand tha paltan officer hai Alpha dilayega. Ab number mile hai chalees percent. Ab Charlie (lowest grading) nahi milega toh aur kya hoga. Ham sale kauwye hain, ham kabutar banne ki koshish kyon karte hain. Ham sale kauwye hain, chahe kuch bhee karen kauwye hee rahenge,"* and went to sleep again covering his face with the pillow.

(Lying in his frenchies, he turned around and said, "What get up? Like a clot I was talking big. I was flaunting that I had brought PCK of an Alpha type, that my unit officer was an instructor here. I was saying that I had come here to get an Alpha and let me see who can deny it to me. And now I have got forty percent marks. With these goddamn marks, I deserve a bloody 'Charlie' (lowest grading), what else do I deserve? Tell me, do I deserve anything better? No I don't. I am a bloody crow, why should I try to become a pigeon. I am a bloody crow and whatever I do, I will always remain a bloody crow," and went to sleep again).

I felt sorry for Pradhan. His wonderful world had come crashing down in no time. But then, he was relaxed the next day as if the monkey was off his back. He had accepted his fate. Thereafter he was found more in the bar and less on the study table, as did most of the course after the midterm test. He then told me that he had decided to quit the army and migrate to USA. Finally, despite all support infrastructure he landed with the omnipresent 'Charlie'. I am not sure if the JC results had anything to do with that decision, but he did actually migrate to the States.

Papa, How was The Course

JC Course leaves more broken hearts than happy ones. In a course of almost two hundred and fifty, no more than ten to fifteen get an Alpha grading. Some forty to fifty get a Bravo and the rest get the 'officer like' grading- the Charlie.

Captain Gaurav was a sincere and a pragmatic officer. A Short Service (SS) Commission officer himself, he was very frank about what he thought of SS officers, which was not very high. Doing the JC course, he would say that a SS officer tries everything in life, the NDA, IMA and all other avenues in the civil. And when he gets nowhere, he comes through the SS commission. The army takes him in because of the shortage of officers, and at worst, they only have to suffer him for five years and then off he goes.

He had coined a word for the SS entry, the 'Shiv Sainiks.' Seeing Padma Kumar, another Shiv Sainik toiling too hard, he would tell him not to waste his time and work according to his potential. Knowing Padma's past accomplishments, I thought it was a very practical advice. Nevertheless, no harm working hard. No one has ever died of hard work, so one can take a chance.

As mentioned, I had landed for the course without pamphlets. He was kind enough to share his with me at his own expense, which meant he had to manage his time to accommodate me, and that put undue strain on him. After the first written test, which more or less decides the officers' grading, he came to my room. I asked him how he had fared in the first test. He was a bit dejected and narrated his feelings thus:

"Look yaar, I am a Shiv Sainik. I tried to join the army through NDA, tried for IMA and tried everywhere in civil but when I could not get anywhere, I joined the Army through the SS. I understand my capabilities and limitations and have never aspired for too much. On this course I was only aiming for a Bravo. I thought if I work hard, and with a bit of luck, I may get a B(I) (Instructor grading). Today they gave us the written marks, and I have got forty percent in the written test. With these marks I can at best only get a 'Charlie'."

"Even 'Charlie' is fine, have got some earlier also. The problem is that my kids have now grown up. They know that I have come for a course and that on the courses there is grading. When I get back, they will ask me papa how was the course? I will tell them it went off well. They will then ask me what grading have I got. All my life I have told them to work hard and always tell the truth. I cannot even lie to them. Then

I will have to tell them that I have got a 'Charlie'. They won't say anything but I know they will think, *Kaisa Charlie baap mila hai, phir 'Charlie' le ke a gaya.* That is my worry, grading is fine." (What a Charlie of a father I have got, is back with another 'Charlie'. That is my worry).

A genuinely good and sincere man who had worked hard. I felt sorry for him.

The Desolate Bar

During the beginning of the JC course, Major Shyam Kumar of 5 GUARDS (later a three star General), went to the student officers' mess bar and ordered a drink. Shyam Kumar with a wonderful attitude was a sharp flamboyant fellow who mixed work and fun well. Normally, the bars are a crowded place during the courses. Noticing that the bar remained more or less empty throughout the evening and no one else came in for a drink, he asked the civilian bartender who had been there for decades.

"Kyon bare bhai, Hindustan fauj ke afsaron ko kya ho gaya hai? Koi bar mein najar nahi aah raha hai." (What is wrong with the officers of the Indian Army, I don't see anyone at the bar).

"Saab pahla written test ho jane dow, phir dekho bar mein khare hone ki jagah nahi milegi," replied the barman. (Let the first written test be over saab, you won't find a place to stand here).

This was after his long experience in the mess, where he had served for more than fifteen years. Everyone thinks he is doing bloody well till the first test. And then, after the written test, everyone knows where they stand, and most stand next to the bar, for the rest of the course.

Gonde Won't Give up

As I said, the JC Course is a do or die effort for some, and they would go to any lengths to get a good grading. Before I go further, a word on the assessment system would be in order. The assessment is based on two aspects. The first is the written test performance (two tests - first after four weeks and the final test towards the end of the course) and the second is the instructor's assessment marks based on performance in class discussions. There is, therefore, always an effort on part of the student officers to impress the syndicate instructor, the senior instructor (SI), a Colonel, and the chief instructor (CI), a Brigadier, with their tactical wisdom and communication skills. The CI has four divisions to go to, each with six syndicates, thus, attends the syndicate discussions rarely. Occasionally, he attends the central discussion of the divisions where the entire division is present. When that happens, there is a sudden increase in the class participation by the students to grab the opportunity to impress him with their little knowledge.

During one such central discussion in which there were more than eighty students present, the CI entered the hall. Suddenly, the participation by the student officers reached a crescendo. There would be many officers jumping up simultaneously and answering the question. Funnier, the smarter ones would start answering the question while still seated and then get up. It took the SI, who was conducting the class, some effort and skill to establish control and restore semblance of order, which nevertheless broke down every now and then.

I noticed one artillery officer Major Gonde, who had been very serious on the course and well prepared, trying to get up and speak, but without success. Every time he was beaten by someone who was quicker off the starting blocks. Invariably, before he could get up, there would be someone already up. Or, to gain those precious seconds, someone would first start speaking and then get up. Sitting across the hall, I noticed his plight and pointed him out to my neighbour, saying something was about to happen. The CI had been there for a while and was not likely to be there for too long. The time was running out and Gonde was getting desperate with every passing minute. He had to do something, or else, he would miss the opportunity to impress the CI. Sure enough, like a leopard on the prowl he waited and pounced.

The discussion on a particular requirement finished and the SI said,

"Gentlemen, with this we come to the end of this requirement and now move to the next requirement." As he said this and prepared to read the next requirement, Gonde, who had been unsuccessfully trying to speak for long, sprang to his feet. The instructor looked at him and assumed he had some doubts on the previous discussion. He asked, "Yes Gonde, any questions?"

Gonde replied, "No sir, I have no questions, but I will answer your next question." The class including the CI burst out laughing. As the luck would have it, CI had run out of time and did not stay on to listen to Gonde's pearls of wisdom on the next requirement.

Wonder what must have been going on inside Gonde's mind, and how desperate he must have been, to devise this 'out of the box' proactive strategy.

Artillery Fire Plan

Before I narrate this incident, I must explain a few terms which may be new to those not conversant with the military. On this JC course, number of classes are conducted in the form of what is called the TEWT (Tactical Exercise Without Troops). It is simply an outdoor discussion on a given tactical setting. The syndicate is divided into sub-syndicates comprising officers from all arms and services. The sub-syndicates produce their respective plans which are discussed in the class.

An artillery fire support plan is an important part of tactical discussions for which an artillery officer is distributed in each sub-syndicate. Fire plan can be expressed either in a written tabular form with all the details or in a diagrammatic form. In a diagrammatic fire plan important details are written on a simple sketch depicting the objective of attack. Contrary to the impression that the name suggests, it is not an elaborate diagrammatic depiction.

During the course, I was sharing my room with a course mate, an air defence artillery officer commissioned from the ranks. Since artillery is an important part of the course and discussed in every requirement, I was happy that I would benefit from him. Those days air defence artillery was a part of field artillery and wore the same uniform. This notwithstanding, the air defence gunners did not associate themselves with the field gunners, and being entirely a different cup of tea, did not bother to learn much about their business either.

One day I walked into my room after a jog. My friend was at his study table with a rather multi coloured pictorial chart on which he had been working for long. I was surprised that the officer had time on this busy course to indulge in a hobby and do painting. I asked him what he was doing. He replied that in his sub- syndicate he was the only gunner officer and he was making a diagrammatic fire plan for the next day's outdoor exercise. Seeing too much colour on the chart for comfort, I went closer. To my amusement, what do I see?

In the artillery resources we were provided four fighter bomber missions and six artillery batteries. My friend had shown colourful fighter bombers with glowing after-burners, diving towards the objective menacingly, with series of bombs hurtling towards the objectives. Six batteries of artillery were diagrammatically depicted by making eighteen guns lined up in a row with rounds flying and landing on the target with full ferocity. With all the colours thrown in, it looked like Picasso's impression of modern battlefield.

I asked him if he was sure that this was the way to make a diagrammatic fire plan. He said he was a gunner and he knew at least that. Well, I said, "you may like to check that out." He did not. I can only imagine what would have happened in the class the next day. I am sure if that fire plan had fallen in the enemy's hands, he would have vacated the post without a fight.

A Fiasco Kohli Would Never Forget

The JC course had run its full course. There were few rats left in the race. Kohli was one of them. He had worked hard throughout the course. The all important final written test had arrived. We sat in the verandah at our desks as the question papers were distributed. Friend Kohli sat next to me. The most important question carrying 50 marks was based on a tactical situation. We were required to site a company size screen position as part of the battalion defended area. There was a four page long narrative giving out review of the situation starting from corps zone downwards. They gave us four map sheets to go with the mostly irrelevant narrative. The area of interest was in the bottom most square of the last map sheet. A half page cyclostyled sketch of the area was provided to mark our screen deployment.

We spent a lot of our time trying to join the map sheets in the windy conditions and got down to reading the narrative. It is only when one reached the end of the four page narrative that one realized that the crux lay in the last paragraph of the narrative. Well, as far as siting of the screen position is concerned, there were two options. One could either site it on the high ground or base it on the built up area. Sitting next to me was Kohli who had been struggling with the map sheets in the wind. Unable to manage the maps on the table, he chose to sort them out on the floor. He pinned one sheet to the floor with one knee, the second with the other knee and the remaining two sheets with his hands. Realizing he needed free hands to put the paper pins, he literally laid 'spread eagle' on the map sheets to do that. A comic sight alright, he finally managed to stick the paper pins and got the maps under control.

Wrestling bout with the maps over, he got down to the narrative. By the time he had finished reading, he was running grossly out of time. He took out the cyclostyled sketch of the area. Unable to decide whether to deploy the screen in the built up area or the high ground, he put his marker pen on the sketch and looked heavenwards for a clue. That must have been quite a while, for when he looked down, he yelled, "Oh shit". I looked towards him. The sketch was cyclostyled on cheap blotting type paper. While friend Kohli was looking heavenwards seeking divine help, the paper had absorbed whole lot of ink and covered the entire area with one big blue screen. Both options for screen had disappeared without a trace in that blue tsunami.

Kohli was visibly shattered. He sat there for a while absorbing the shock and awe, and then as the time was running out, ran to the instructor with his blotted sketch requesting for another. "Sorry," said the instructor, he did not have one. He ran back and got down to making a sketch himself. It was a case of too little too late. The instructor announced it was time and took away our answer sheets. It was another 'tragicomedy'.

Kohli's world had drowned in the deep blue sea. No wonder, they say there are many a slip between the cup and the lip.

The Defence Services Staff College, Wellington

The Staff College is a tri service institution located in the beautiful Nilgiri hills at Wellington. With rolling mountains, vast tea gardens and an excellent climate, it is a perfect place to spend one year. The curriculum those days was not too hectic with lots of social activities forming part of the calendar. The ladies too had enough to do there with lots of official and social engagements of different kinds.

The Staff College was considered a long paid holiday. Some feel there isn't much to do, others don't agree. The former always base their argument on hard statistics which is hard to refute. It is proved by the fact that during Staff College course, anything up to 100 student officers' wives, out of about 300, conceive in the salubrious and serene environment of Nilgiri hills. The gynaecologists and the paediatricians are nowhere as busy delivering goods as in Wellington. Surely there is enough time to spare!!

Qualifying Staff College is an important landmark for an officer. Everyone works hard and gets into the Staff College after a competitive examination. At that stage everyone believes, once through, it is a passport for higher ranks. It is not necessarily true though, as over sixty percent officers are over looked in the rank of Lieutenant Colonel alone and the genocide continues in all higher ranks, being a vacancy based system. This notwithstanding, once there, everybody sees himself as a future General and some land up with inflated opinion of themselves.

Many student officers keep hinting that they did exceptionally well and were on top of the competitive merit. This they do, as there is no way to find out the truth, as the merit list, if there is one, is never made public. One instructor in the Staff College remarked,

"When I did the Staff College Course in late seventies, I went around saying that I was second on the course. Somebody asked me what makes you so sure that you were second, I replied, because everyone else is saying he was first."

So that is the environment in which the course runs its course. The student officers work hard, and like the other courses, are very keen to do well to be able to get a good posting. In the bargain, many interesting events take place as we shall see.

A Tit for Tat

Even though the entrance to the Staff College is based on competitive examination, there are some surprise arrivals as well. The Staff College gives some people a wrong idea about themselves and develop ambitions beyond potential, not a good news for their would-be-subordinates or for themselves.

Some wives get even more excited. I am told the wives have an occasional friction as to how one's husband was posted as the Brigade Major (BM) and the other's, the DAQMG of a brigade. BM is a General Staff appointment and is considered more glamorous of the two. In an effort to do well at this all important course, some resort to one-upmanship.

Some out of those who finish in the competitive list, that is the top twenty, continue to refer to it till they step into their grave. Those who are not in that list claim they were the twenty first. No one has a way to verify and none is interested.

On the course, the student officers, hereafter referred to

as students, are required to deliver a talk on a subject of their choice. It is followed by a short question answer session. To avoid getting caught with an awkward question and to impress the instructor, some students plant a question which they prepare well.

One officer planted a question for his talk through Major Sohi. As the talk finished, he invited questions from the class. Sohi asked the question as requested.

"Gentlemen, the question asked by Major Sohi is not relevant to the subject and beyond the scope of my talk. Nevertheless, I will try to answer the question," and proceeded to answer the question very well since it was well prepared.

Major Sohi sat there dumb founded not believing what was happening. He had asked the question on his request and was made to look silly. He could not digest that. He stood up and said,

"Sir, it is rare that I ask a question in class." Holding aloft a piece of paper for all to see, "This is the question he gave me in the morning and requested me to ask at the end of his talk. Now he says this question is irrelevant and beyond the scope of his talk. I thought it necessary that I must let this be known."

The instructor and the class were pretty amused. The class complimented Major Sohi for the tit for tat.

All Roads Lead to Lockers

In the Staff College, each officer is allotted a locker with a number. All the instructions, issue of books, exercise papers, private letters and what have you, is done through the respective lockers. Even the instructions to the lady wives about their programme are passed through the officers' lockers. All student officers are required to write their locker numbers in all correspondence and assignments. There is so much insistence on periodic clearance of lockers and mentioning of locker number in all submissions, documents and everything else that locker numbers threaten to replace the names. One day, when an instructor's wife asked a student officer's wife, "And you are?" She replied, "I am Mrs. 420." (her husband's locker number).

What's in an Officer's Name?

Most classes at the Staff College are conducted in the form of tutorial discussions in which the students do all the talking. The instructor guides the discussion. With the aim that everyone gets to know each other, they insisted on you giving out your name and the regiment before speaking, particularly in the beginning of the course. The foreign officers gave out their names and countries. The instructors are extra polite to the foreign officers. Even if they are at a tangent in the discussion, it is tactfully re-worded to sound as if that is what the officer meant to say.

There was an officer from the Middle East. He was called Abdullah. He had a very long name, similar to the crown prince of Saudi Arabia with a few titles which ran into two and a half rows on his rather largish name plate. To top it all, when he spoke English with a heavy Arabic accent, some thought even he did not understand himself.

During a class, Major Abdullah was asked to give his views on a company quick attack in mountains. Abdullah got up, collected his thoughts, spoke for a while and paused. The instructor having listened to him carefully said,

"Thank you, please sit down. Gentlemen, as rightly brought out by Major Abdullah ... ," and before the instructor could say anything further, Abdullah interrupted.

"Sorry sir, so far I have brought out nothing. I have only given out my name, country and locker number."

It's Lunch Time so..

In my division in the Staff College, the Senior Instructor (SI) had a typical habit of summing up the class well into the lunch time. The students obviously resented this but could do precious little. Just to convey their dissent, students would start looking at their watches and watch other students walking past their class. You could hear a sigh of relief the day he did not do it. One day, the discussion dragged well into the lunch time because of some 'Keen Kumars'.

When the students thought the class was finally over, to everyone's disgust, the SI walked over to the rostrum for the last word. Realizing that it was well into the lunch time and the disinterest of the class was oozing out from all pores, he said,

"I know gentlemen that we have over shot the time. And since it is lunch time and you must be hungry, I thought I must give you some food for thought," and proceeded to give a long summing up discourse. 'Food for thought' at that hour left no appetite either for thought or for the stomach. A good sense of humour that did not amuse many.

Logical Corollary

Officers on courses are required to bring equipment like binoculars, compass and map cases from their units. Always worried about losing them and getting into trouble, many officers avoid bringing them to the course. Those who do bring them prefer to lock them up in the box and not carry them for the exercises.

During an outdoor exercise in Staff College, the discussion was based on defence in mountains. The instructor, Colonel Nirbhay Sharma, a paratrooper, asked Major Balwan, a short stocky surd to nominate the 'Ground of Tactical Importance'. Finding difficulty in indicating the feature on ground, the instructor asked Balwan to show it on the map. Balwan who was not carrying his own map started looking around for one. Seeing this, Colonel Sharma commented, "Balwan, I notice you have started travelling very light these days."

Balwan borrowed a map and showed the hill he had nominated as the 'Ground of Tactical Importance'. When asked for his reasons for the decision, Balwan said that it was the highest mountain in the area. Colonel Sharma remarked, "If India was to go by your logic, the defence minister of India would have nominated Mount Everest as the national 'Ground of Tactical Importance'.

Self Contained for Seven Days

We were going through a war game in the Staff College based on offensive operations in high altitude terrain. At that altitude employment of helicopters for resupply was ruled out. Further, it was impossible to move self contained for anything more than three days because it amounted to carrying impossible loads. The operation itself needed seven days to complete. Left with no option, the brigade commander planned a manoeuvre based on troops self contained for seven days, hoping that no one would notice it.

Obviously that did not get past the instructor. He asked to be explained how the troops were self contained for seven days and how much was the load each soldier was carrying. The brigade commander knew he had planned only five days rations and would be caught. He got up confidently, explained that his troops were self contained for seven days as follows:

"Sir," he replied, "Three days man pack, two days re-supply on mule loads and .... two days hungry."

Soon an arty shell landed on his brigade headquarters and he was declared dead. His deputy commander, the inimitable Major Balwan was asked to take over command.

Sand model exercises in the courses tend to become a sham for students who do not hold important exercise appointments. Such students become passengers and watch the fun from the sidelines. Deputy brigade commander in exercises, as in real life, has a different charter of duties, and invariably tends to be left out of 'matters operational'. During the courses the deputy commanders sit back, pretty certain that they have no role to play in the proceedings. Balwan was making good use of time to catch up with some sleep from the back bench. When occasionally awake, he was having a great time watching the appointments getting grilled. Much to his unpleasant surprise, Balwan was unexpectedly caught in the thick of battle when the brigade commander was declared dead.

True to his appointment, he had not kept himself abreast with the progress of the battle. When he force landed on the sand model to explain the situation, he found himself all at sea in the rugged high altitude terrain. He was moving the pointer rapidly all over the place, waiting to be prompted by someone as to which hill feature was under attack. The SI watched for a while and chipped in, "Balwan, slowly my friend, slowly. You are an infantry officer. It is a rugged high altitude terrain, over 14000 feet, and in this terrain even the pointers move slowly."

When Balwan, aided by prompting, finally located the mountain under attack, he gave an ambitious time and distance appreciation. Not convinced, the SI asked for his rationale. Balwan said, "The crow flight distance to the objective is ten kilometres and at two and half kilometres to an hour, it would take about four hours to get to the objective."

The SI, as earlier, butted in, "Balwan, I told you that it is a very high altitude terrain and like the pointer, in this terrain, even the crows do not fly straight. You may like to reconsider your time and space."

Balwan was quite delighted to be appointed deputy commander, hoping to catch up with some sleep in the sand model room. He had never imagined he would find himself in the thick of battle at 14000 feet.

Commandant and his Golf

The Commandant of the Staff College, Wellington in 1990-91 was neither particularly good, nor regular at golf. He was a very humorous person though. He said the joke going around the Staff College about him was, quote "What is the handicap of the Commandant in golf? And some one would reply, his golf, of course. And when do you know that the Commandant is on the golf course? When you find that even his dog is hiding under a car," unquote.

Taking out Sand from Shit

Here are the extracts of a lecture taken by a student during the Staff College course on exploiting natural resources for energy in India, clearly demonstrating the originality of thought.

"Talking of the natural resources for generating heat energy, I would take the example of human wastes. India has a population of about one point one billion. Taking on an average that each man shits one kilo daily, India produces one billion kilos of shit every day. One kilo of shit can generate xxxx calories of heat, which means that if India can harness sixty percent of its human waste resources, it can generate over xxxx calories of heat, equivalent of xxxx megawatts of electricity. I would now highlight the progress of 'Gobar gas plant' (plant to generate cooking gas from animal wastes), a very successful venture in rural India. India has an estimated xxxx million cattle heads. Considering each cattle shits two kilos per day & sixty percent of it can be harnessed, we should have xxxx thousand tons of shit per day. So this amount of shit will generate xxxx megawatts equivalent of heat energy. The human and animal shit put together can help India tide over the energy problem in rural areas for good.

What needs to be appreciated is that it is biodegradable, pollution free and a renewable source of energy with assured re-supply requiring no additional investment."

It was a thought provoking lecture. I was not in that class. I am told the class found it amusing and realized how much shit was flying around in India, all going down the drain. A bit unconventional and out of the box alright.

Ignorance is Bliss

Contrary to general belief and expectations of officers who join the army, there is a lot of studying to be done in the armed forces. Officers are required to pass promotion examinations for almost every rank. A number of officers do not take the promotion examinations seriously till the water starts flowing over their heads. For promotion from Captain to Major, the officers are required to pass promotion examination Part D. The examination is held once every year and an officer gets about five chances to pass five papers. Even one paper at a time is permissible. All in all, not a difficult proposition if an officer takes it seriously. Unfortunately, some officers feel there is enough time, and in the bargain land up in a tight corner.

In the Tactics paper of one such examination, question number 5 required the officers to draw the outline organization of a Parachute battalion. Capt Negi of my unit was taking the test and sitting behind him was another infantry officer. The officer behind did not know the organization of the Parachute battalion. Seeing that the invigilator had gone out of the hall, he took the opportunity and asked Capt Negi for a cue.

"Yaar question number 5, organization of Parachute battalion." (Partner question number 5, organization of a

Parachute battalion).

Negi: (Partially looking back, realizing that it is impossible to dictate the full organization of the battalion), "It is exactly the same as infantry battalion."

"Chal phir woh hee bata de," (OK, tell me that then), said the infantry officer behind.

Siting the LMG

- *JCOs/NCOs – Junior Commissioned Officers and Non Commissioned Officers. They are the backbone of any army.*

During the courses some students master the art of giving their views without committing on any issue. They believe, as long as you remain vague and do not commit, you always retain the flexibility of manoeuvring out of a corner. "Oh, I meant this and not that." This art also percolates down to the JCOs and NCOs who also do various training courses.

The army conducts Platoon Commanders course for JCOs/NCOs, where they are trained to lead a platoon in war and peace. This course puts the students through some tactical discussions. One such exercise was on occupation of defence by a platoon in plains. The class was explained the tactical situation by the officer instructor. The students then went around to reconnoitre and study the ground. Thereafter they sat on the ground for discussion. The instructor asked the class, "In the given situation as explained, where will you site the Light Machine Gun (LMG)?"

One 'Keen Kumar' Sikh JCO from Sikh Regiment got up, made some general vague statements on factors to be considered while siting the LMG and sat down.

The instructor asked again, *"Saab maine poocha hai, zamin par batao LMG kahan lagaoge?"* (I have asked you to tell me where exactly will you site the LMG on ground?)

The JCO said, *"Jahan tak LMG lagane ka sawal hai, woh toh zamin ki banawat aur larai ki halaat par nirbhar karega."* (As far as siting of the LMG is concerned, it will depend upon the ground condition and the tactical situation). The instructor, himself a surd (Sikh), had - had enough of such nonsense and said,

"Saab, larai ki halaat toh main aap ko pahle bata chukka hoon aur zamin ki banawat aap ke bund de thalle hai. Hun dasso LMG kithe lagegi?" (Saab, the tactical situation, I have already explained to you and the ground situation is right under your arse. Now tell me where will you site the LMG?)

There was stunned silence in the class. That was the last time anyone tried that stuff on that course.

To Err is Human, to Blame it on Others...

Soldiers, like the officers, attend courses in various schools of instruction. Every jawan who returns from a course of instruction tells the story of how hard he worked, how well he did and how much he deserved 'Alpha' grading and how he missed it because of unfair treatment.

Guardsman Gopal of my unit, a Garhwali, had returned from a course. He was being interviewed by his company commander Major Negi. It proceeded like this:

"Haan bhai Gopal course kaisa raha?" (How was the course Gopal?) asked Negi.

"Theek raha saab," (Went off well), replied Gopal.

"Kya grading milne ka umeed hai?' (What grading do you expect?)

"Mid- term tak toh Alpha chal raha tha saab." (Was running in Alpha till the mid- term break).

"Aur mid- term ke baad kya chal raha tha." (And after mid-term?)

"Waise toh Bravo mil jana chahie, lekin Charlie toh kahi nahi gaya." (Should get a Bravo, but no one can deny me a Charlie for sure).

Major Negi then said, "Mid- term ke baad tera panga ho gaya hoga." (Must have got into trouble with the instructor for no fault of yours).

"Haan saab." (Yes sir).

"Wahan kisi ne paltan ke khilaf kuch bol diya hoga, tere se suna nahi gaya hoga aur tu ne jawab de diya hoga aur usne teri grading kharab kar dee hogi." (Someone must have said something about the battalion, and you must have fought over it, and he would have spoilt your grading).

"Haan saab." (Yes sir). And then realizing he was falling into a trap *"Nahi saab."* (Yes sir, No sir).

"Nahi toh mid- term test ke baad tu zarur bimaar ho gaya hoga." (If not, then you must have fallen sick after the mid-term).

"Haan saab, mid- term ke baad pura course taklif rahi, is liye mehnat nahi kar paya." (Yes sir, I remained sick for most part of the course after the mid- term, so could not work hard).

"Aur Charlie le aya," (And brought Charlie), said Major Negi

"Haan saab, aap ko toh malum hai waise wahan bhai bandi toh hoti hee hai. Apne group ka toh wahan koi tha hee nahin, is liye hamari grading kahan ban nee thee," replied Havaldar Gopal.

(Yes sir.You know there is so much of favouritism there. We had no one from our regiment, so there was no chance of getting a good grading).

"Ab mujhe samajh aya ki aap ki Alpha grading kyon nahin ayee. Agli baar beemar mat hona, panga mat lena aur insha allah, apnee regiment ka koi instructor wahan hoga toh aapkee achee grading ayegee. Tab toh layega na?"

(Yes, I understand why you did not get good grading this time. Next time do not fall sick, do not get into trouble in the name of the regiment and by God's grace there will be an instructor from our regiment and then hopefully you will get Alpha grading, will you?) said Negi.

"Pakka lounga saab," (Then I will definitely get Alpha grading), replied Gopal.

Later, Gopal went for another course with the same result, returning with another Charlie. He was luckier this time though, his company commander Major Negi had been posted out.

This is roughly the story of most who get a Charlie grading, jawan or an officer. Those who do well give no excuses and do well in every course that they do.

Memoirs from Senior Command Course

The team at work

Senior Command course conducted in Mhow trains potential commanding officers in essentials of unit command. The officers from all arms and services attend the course in mixed syndicates. The course has only two gradings, Qualified (Q) and Qualified (Instructor) or Q(I). It is, thus, a laid back course with a relaxed atmosphere. There is enough time for tennis and golf and hanging around. There is no rat race unlike other courses where student officers tend to go overboard in pursuit of Alpha grading. On the flip side, many take things rather lightly.

Mostly, it is the officers from combat arms who work towards getting a Q (I). Those from the services, by and large, are happy going through the motions and do not trouble either the instructors or themselves.

The course is divided into syndicates of about ten student officers each. The classes are mostly conducted as syndicate discussions and through outdoor exercises called TEWTs (Tactical Exercise Without Troops). For TEWTs, a tactical situation is given to the class. The sub-syndicates within the syndicate are required to prepare a plan, mark it on a sketch and submit it with answers to the requirement.

The sub-syndicates meet in the afternoon for formulation of the plan. It is here that the degree of interest and contribution of an officer in the course comes alive. In my syndicate it was like this:

Lieutenant Colonel Pitambar, Signals - "What time are we getting together yaar? Four p.m., is it ok if I come by five thirty? Just do not worry, I will come and deposit the file."

Lieutenant Colonel Santa Singh, Army Service Corps - "I am volunteering to collect and deposit the narrative from the DS co-ord. Don't look for me during the discussion." He would come sharp at the pre-decided time, hand over the whites, confirm the time for depositing the syndicate plan and disappear. He would then appear at pre-determined time and deposit the work.

Lieutenant Colonel Dhingra, Ordnance- "Give me the file cover yaar. I'll prepare the file cover and write the syndicate names." That was his contribution of the day, and would disappear for good.

Lieutenant Colonel Sidhu, Gunner- Always on time to begin with, "Tell me what time do you want tea, yaar. What else, want samosas also?" He would then vanish and return with tea at the given time and disappear again. He would again appear a little before the deadline for depositing the file to make the necessary notes to survive the next day.

Lieutenant Colonel Arjun Dallal – "Give me the sketch yaar, I will highlight the roads and the rivers on the sketch. I will leave my sketch pens here. Sengar, you may need them. Remember to return them to me tomorrow." And having done a decent job, "Anything else I can help with?" And without waiting for a reply, "Nothing, then see you tomorrow. Just tell me the major points of the plan tomorrow in the vehicle. Goodnight gentlemen."

There were other 'Keen Kumars' who believed in economy of effort yet wanted to do well. They would appear in the last half an hour, "Just tell me the plan for tomorrow yaar." They would pick up the plan and quietly make detailed notes and disappear. They would be most active in the class the next day, and pose as if they were the main architects of the syndicate plan, if it was received well. If not, they would be silent observers as if they had nothing to do with it.

Pitambar, true to his word, would appear just in time, put the file together, wish us goodnight and deposit it with the duty clerk. Somewhere on the way, he would make a few notes on the plans to survive the next day.

The other gang members who contributed with tea and samosa, highlighting and preparation of file, would encash the first opportunity to speak during the discussion, to tick mark as having 'Spoken for the day'. They would then go into deep hibernation unless disturbed by the DS.

Great course - 'officer like' as it is called. These wonderful folks made sure we enjoyed our stay at the Army War College.

Fear of Getting Exposed

Bajwa is a short-statured, cheerful sweet fellow from the JAT Regiment. Despite his fifty plus age, he still has a baby face. During the Senior Command Course, he would keep us amused with his anecdotes. During the course, there were three performance counselling by the Chief Instructor (CI) to apprise the students how they were faring. For this we would wait in the syndicate rooms and went to the CI by turn. The CI would tell us how we were performing and where we needed to work harder.

I used to call Bajwa 'Colonel'. After his counselling, I would ask him "How was it Colonel?" He would give an emphatic thumbs up and produce a funny chuckle, 'trrrrr....' "Going great, running for instructor grading," he would reply with his typical smile.

At the end of the course, the final counselling was conducted by the CI and the course gradings were also being conveyed. As the Colonel returned after his counselling, I asked him, "Final counselling, how did it go Colonel?"

This time he surprised me with an emphatic 'thumbs down' and even a funnier chuckle 'trrrrrrr....' and said, "All is in the well."

Well, we all went back to our units from the course. Both our battalions were in the same formation. I met him one day in Ganganagar as he was walking to his unit from home.

I asked him what was up. He said he was caught in a dilemma and was figuring out what to do. "You know, my dilemma is that I have been nominated to attend Combat Group Commander's Course. It is a command course for mechanized and armoured officers. The fact that I have been nominated after having done the Senior Command Course shows that I have also become an upcoming officer and am doing well. The problem is, there is a formation exercise coming up during the period of the course. Since I've done Senior Command Course, my commander has asked my commanding officer to take up a case for cancellation of this course.

"Sad," I said. "You are missing an opportunity to do a good course."

"Oh yaar, that is hardly a problem. The problem is that for the first time in my life, my commanding officer wants to give me an outstanding report. For this, the brigade commander is to be informed three months in advance and he would interview me. If I don't go on the course, he will interview me and I'll get exposed, *aur lene ke dene par jaenge aur kahin eight point report bhee na khatre mein par jae,"* (May land up as a loser and might endanger even my eight point report), he said with his mischievous smile. "That's why I need to go on the course."

I am not sure what finally happened about that course. He later commanded a battalion with distinction during the Kargil war and was awarded Vir Chakra. Sitting in those tough conditions waiting to attack the impossible heights, he would receive *"we are with you"* messages from some regimental officers sitting in Bangalore or such far off places. Sitting amidst exploding artillery shells, he would joke, "Oh son of a gun, if you are with us, come here and join us. Digital support from Bangalore is too distant and too useless."

Bajwa was great fun and could laugh at himself anytime, anywhere.

BEYOND THE SEVEN SEAS

Recalling UN Mission in Cambodia

I had the opportunity of serving with the UN Mission in Cambodia in 1992-93, called UNTAC. Cambodia is a unique country in many ways. Devastated by the reign of brutal Pol Pot's Khmer Rouge, it was devoid of any worthwhile industry. It was characterized by lush green forests, with the Mekong river meandering quietly right across the country and virgin white sand beaches. It was a country as close to nature as one can get. Tonle Sap is a huge lake that expands many times its size in monsoon. People lived on boats and interestingly maintained floating kitchen gardens and small fields. They towed these fields behind their power driven houseboats as they moved from one place to another. It was an ingenious way to combat devastating monsoons that often hit Cambodia. Flying over the country in a helicopter, one could see the wonderful forests dotted by small thatched hamlets. The terrain was pockmarked by huge craters, reminder of the carpet bombing by the Americans during their war with Vietnam.

Cambodians suffered greatly from the mines laid during the civil war. The poor Cambodians, barely making two ends meet, would be smiling all the time taking difficult life in their strides. Large number of mine casualties with serious amputations, some with all four limbs gone, would be begging, but the spirit and the smile was there.

The arrival of UN Mission, called United Nations Transitional Authority in Cambodia, UNTAC, had triggered a construction boom. I had rented a part of a house for myself, and for some time, my family joined me in Phnom Penh. Standing in the balcony of the house, I saw a glimpse of a weapon in the attic which looked like anti tank RPG 7. Interesting, a rocket launcher in the house! I decided to ask the house owner.

"What is that in the attic?"

"It is RPG, anti tank weapon," he replied.

"What is it doing in your house?"

"My personal, I have kept it for my protection," he said.

Great place I thought, people having personal anti tank weapons. Not surprising, in Cambodia, marriages were celebrated by gun fire and every household seemed to have enough weapons for all round defence.

My two sons were small, two and three and half years respectively and spoke decent English. The house owner was clearly impressed. "Where did they learn English?" he asked us. We told him, they learnt it at home as we spoke in English in front of them and that's how they picked it up naturally. A little later he was blessed with a son. I was quite amused to see that he bought English music cassettes, Madonna and Michael Jackson included, and would play them next to his son's cradle who was less than four weeks old. Good ambition and foresight at its best.

It was an experience to see the UN at work. Coming from the Indian Army where austerity is the norm, it was some change. Well, back in units we use a sheet of paper till there is not even a square inch of blank space left on either side.

There were more computers in the office than the number of officers. The furniture was simply top of the rack. It made life comfortable but did not add to the efficiency of the staff. In fact, it was an avoidable extravaganza.

Working at my state of the art desk, I had hung a wad of used A4 sheets. A Canadian, seeing them, asked me, "Why are you piling up this crap, mate?"

I said, "I will use the unused side of the paper for drafting and rough work." The Canadian was amused but clearly appreciated the idea and started doing it himself.

The stay in Cambodia too was full of fun. The situation in

the country then was warlike. The UN was there to implement a truce between the government and the deadly Khmer Rouge. One major problem in Cambodia was that of mines. It was common to see people, young and old, with amputated limbs almost everywhere. Helping them overcome this problem was one of the UN mandates.

Here Destroy These Mines

One day the Chief of the Engineer force in UN, a Colonel from New Zealand, was surprised to find a young Cambodian boy walk into his plush office with a live mine in his left hand and a mortar bomb in his right. The boy just walked up to him and said in Khmer, "Here-destroy these."

With an army of about five thousand, New Zealand has not fought a war in decades. The biggest Engineer unit I am told is a company. The Colonel would not touch the bombshell with a barge pole. He had never faced a professional challenge like this ever before; a live mine and a blind bomb in his office.

The Grass on the Other Side of the Fence

The United Nations operations are characterized by multinational nature of its functioning, sometimes to its detriment. Even a small group of three would invariably have officers from three different countries. There is generally a problem of communication as not all of them speak English or French, the two functional languages. Some countries tide over this problem by sending interpreters with their contingents.

The initial Russian contingent of Engineers with UNTAC had sent an interpreter who was the only English speaking officer in the group of five. He would be seen hanging around with his Colonel all the time. An Indian officer during a casual conversation with him said.

"Poly, it must be great fun being an interpreter. You have no responsibility, simply translate, that's it."

Poly replied, "To the contrary, it is much easier being an Engineer officer or anyone else than being an interpreter. Look at me, every time a non English speaking Russian officer is on duty, I also have to be in the operations room. I man the phone, read the message and reply the message while the guy actually on duty is sleeping away to glory. I am here every second day, whereas their turn comes once in weeks." So much for a man without responsibility.

Try Dodging a Mess Bill

The general impression in the army is that no good letter reaches you on time and many do not reach you at all. On the contrary, letters carrying mess bills or any other bills will find you no matter where you are, even if the address is wrong. To prove the point, I must narrate this experience.

I was in Phnom Penh, Cambodia, with a UN Mission for over a year. I received two letters coming through the civil postal system, which hardly existed in Cambodia at that time. One was from Staff College, Wellington, which I had left more than a year ago, and the other from the Battle Honours Officers' Mess (Army Headquarters mess), Delhi, where I had stayed over six months ago. Each of these letters was a mess bill of exactly twenty four rupees each. The letter from Staff College, Wellington, had been sent to my unit in J&K, re-directed to Army HQ (UN Cell) Delhi, re-directed again at my home leave address, not found me there- was returned back to Army Headquarters, UN Cell. It then reached UN Headquarter, New York, was re-directed to UNTAC, Cambodia, and finally delivered to me by a room cleaner who had no business to be handling the mail. God only knows how much money was spent on the postage for a Rs. 24 mess bill.

The letter from Battle Honours Officers' Mess comparatively did only half marathon. From this mess in Delhi to my battalion in Kashmir, re-directed from the battalion to Army HQ UN Cell, further re-directed from Army HQ to UNTAC, Phnom Penh, Cambodia. My predicament was how to pay. I had no bank account, no cheque books of Indian bank where I had my account. Even if I had a way of sending a cheque from there, I would be spending more on the stamps and outstation charges than the bill itself. What would you have done under the circumstances?

More often than not, the officers, rightly or wrongly, do

not agree with their mess bills. There are three kinds of officers when it comes to dealing with the mess bills. Very few clear them straight away, an endangered species nearly on the verge of extinction. Some put it in 'pending' tray to be attended to later when free, not that they are busy then. And then there is a category that mark their mess bills NFA (No further action) immediately, an over breeding category. Not that they can get away, just feel uncomfortable clearing the bill promptly.

As for me, under the circumstances, I took a leaf from their book and marked it NFA. But they never gave up and the bills came again when I returned back to India.

I am an Observer and I am Observing

There is a category of officers on UN mission called observers. They are part of all UN missions. They are expected to observe, monitor and ensure that the agreement is being correctly implemented by all parties. On ground they mostly have little to do, for most missions drag on and on. For us, men in uniform, it is frustrating as we are used to doing things fast.

In UNTAC, Cambodia, I was posted on a staff appointment. It was my duty to oversee the operations of observer teams. I visited one observer team in a remote part of the country. An observer team was deployed there to coordinate all affairs of Cambodian Army soldiers who were to be cantoned there.

The team consisted of three members. The team leader spoke only Russian, second spoke only French and the third spoke only Arabic. A difficult situation indeed. To top it all, the locals spoke only Khmer. On seeing the UN helicopter, all of them zipped in the UN Toyota Land cruiser and met us at the helipad. Two wore sleeveless vests, shorts and slippers. One, the more disciplined of the lot wore a striped short, a casual half shirt and a UN cap. I asked them, "Can you take us to the faction cantonment site?"

"What is that supposed to be?" replied the French speaking officer in a hard to understand English.

"There is supposed to be a cantonment site right here. The Cambodian soldiers are to be stationed in it once the agreement is implemented and you are supposed to be responsible for that."

"No one told us anything."

"How long have you been here?" I enquired.

"Fifteen days," he said.

"And what have you been doing all this while?"

"We are observers. We were told to observe, so we are observing, observing and observing," he said enthusiastically in his barely understandable English.

"And what have you observed so far?" I asked.

"Nothing," he said.

Long live UN missions. They do not need the blessings, they are too long anyway and how much is achieved is anybody's guess.

The MSA Crisis

The officers on UN mission are paid Mission Subsistence Allowance (MSA) by the Finance Section to cover all their expenses of food and accommodation. They are on their own for everything. The allowance is handsome, allowing reasonable savings and that is the attraction for serving on a UN mission. During the UNTAC mission in Cambodia, the UN authorities decided that the officers on staff appointments would be provided hotel accommodation and paid a fixed sum as food allowance. Their MSA would be accordingly reduced, whilst the MSA of the observers was to be increased. There was great resentment amongst the affected officers and a flurry to get their status changed to observers.

One officer from an African contingent whose status as an observer or staff was doubtful, left the UNTAC Headquarters without permission and deployed himself with an observer team. A humorous character, his answer to any question, including "How are you?" would be, "I am an observer," and ask him, "What are you observing?" He would reply, "I am observing the Finance Section at the UNTAC Headquarters."

UNTAC Asked For It

When the MSA crisis was looming large in UNTAC, there was much resentment. Life seemed to have lost all its fun for officers on staff who were the affected lot. The staff, when proceeding on duty out of the place of work, produced the bills for hotel accommodation, transport and whatever the duty entailed and that was paid by the Finance Section.

One day the Chief Finance Officer, a civilian, who was considered the man behind stopping the staff officers' MSA, was approached by two young prostitutes. They said,

"Two UNTAC persons used our services for the night but did not pay us. Instead, they gave us a handwritten bill for twenty dollars and asked us to collect the money from the Finance Officer. They said, he pays us only the meagre pocket money, for everything else, we produce the bills and he pays for them. Go get it from him. Please give us our money." The Chief Finance officer was stumped for once.

Looks Can be Deceptive

One of the unmistakable characteristics of the capital city Phnom Penh, Cambodia, were the dancing restaurants. The city comes to life at night. Most restaurants had lively dance floors with blaring live music and singers.

The restaurants were teeming with girls decked up with cheap cosmetics and dressed to kill. The dancing restaurants got a major push with the arrival of UNTAC as girls from neighbouring countries flooded Phnom Penh in search of quick bucks. The UNTAC people returning from remote areas, devoid of life, found these places a good change with their noisy yet lively ambience. These girls would give company to their wards over a beer, dance with them, and were invariably game for more.

One UNTAC man with a nice beard danced away merrily with a pretty looking female, frequently attempting to communicate with the help of gestures as she knew no English. Apparently, after much effort they seemed to understand each other and left the restaurant hand in hand.

A girl giving company to another UNTAC partner nearby kept giggling on their conversation. She burst out laughing as they left the restaurant holding hands. Her partner asked her what the matter was. She replied, the girl with whom he has been dancing and has now left with- is a boy. Human beings continue to surprise me with their ingenuity.

Sharma's Brilliance

United Nations' luxurious way of living was a revelation to Indian contingent in Cambodia . In the offices, every officer was given a state of the art computer and a printer, whether he needed one or not or knew how to use it. Most of the Indian officers were getting to use personal computers for the first time. With little to do, the Indian contingent learnt to play solitaire and chess on the computer.

Captain Sharma from ASC was finding himself grossly underemployed and spent most of his time playing chess on the computer. Impressed by his long sitting at the computer, which conveyed an impression that he was working hard, he was issued a lap top computer which he always carried, not knowing what to do with it. One day he narrated me this story:

Quote, "Sir, you know, yesterday night I was playing chess against the computer. It was a very competitive game. When it was the computer's turn the thinking man would appear till the computer made its move. Once, after I had made my move, the thinking man emerged again as the computer contemplated its next move. Ten minutes on, the computer was still contemplating as the thinking man sat patiently. I complimented myself for the deadly move that seemed to have got the computer in a fix.

Half an hour later I was still waiting, thrilled that the computer had no answer to my brilliance. As I sat gazing at the computer, my neighbour walked in. He asked, "What's happening, you are up so late?" I told him of my brilliant move which had got the computer in a jam for the last half an hour. A little more knowledgeable than me, he pressed a key on the keyboard, nothing happened. He pressed some more keys and nothing moved.

"You idiot," he said, "It is not your brilliance. The computer had hanged half an hour ago."

"So much for my brilliance," unquote, he narrated with wonderful jest.

The Great Interpreters

The UNTAC deployment in Cambodia created a serious problem of interpreters. Very few Cambodians spoke English, some spoke French. The Human Resource Office was hard put to recruit suitable English interpreters. Some recruited were pretty awful, but had to be suffered due to scarcity. The interpreters on the other hand had to cope with almost forty different world wide accents which made things worse.

During a visit to a cantonment site by an Engineer team, one Indian officer asked the interpreter to check with the Cambodian officer when would they start for the next cantonment site. The interpreter, after a long discussion with the Cambodian officer said, your lunch has been arranged in the restaurant.

Entertainment Gratis in the Embassy

I was posted as Defence Attaché in Turkey. The functioning of the mission provided wonderful moments of fun. I would rather not mention all of them here. However, one incident that kept everyone entertained at the mission is narrated below.

Embassies represent the country abroad. As an army officer brought up on the best values of work culture, camaraderie and self sacrifice, one often feels out of place there. Their functioning can both be a source of disgust and entertainment. There are many interesting cases that I experienced. Let me narrate you one.

A lady Second Secretary (PA) had one kidney. She served in US, Turkey and Australia in succession, in all, over eleven years at a stretch. Her monthly medical bill was a fortune, but who cares. When it comes to work, she grabs an office on the second floor. That's where the office of the Deputy Head of Mission was also located. Her boss, the Head of Chancery, is on the first floor. The secretary of the Deputy Head of the Mission, on the other hand is on the ground floor. So as you can see, no one is where they should be. The Ambassador simply does not find it worth addressing the problem for reasons best known to him. She refuses to come down to take instructions or dictation citing her single kidney. But she is happy to climb up to the second floor because she wants to be with the official ISD phone, away from others. So her boss, the Head of Chancery, either types the letter himself or walks up to his secretary on the second floor to deliver and collect the draft.

When the Deputy Head of the Mission on the top floor has to get a draft typed, he phones his secretary on the ground floor. The secretary, not keen to take the long walk up the staircase, declines and waits on the ground floor lobby. The Deputy Head of Mission tosses the paper that flies down amusingly like a 'katti patang'. I thought they must be given credit for use of natural sources of energy i.e. gravity. An entertainment 'gratis', it was like an organized street play as the locals watched with amusement. So much for our missions abroad.

Say hello to...

Gurjit was a medium level functionary in the administrative wing of the embassy in Turkey. I had helped him overcome some major problems relating to pay and allowances of his wife working in the Air Force Accounts office, Dehradun. He remained in touch with me after I had left Turkey.

One day I received a mail from him reading:

"Good morning, sir. All is very well here. You will be glad to know that I am fine, my family is also fine. By your good wishes I have been promoted and posted as consul in USA. On behalf of my family, I extend my warm regards. Please also convey my warm regards to all Tom, Dick and Harry working with you."

Little Knowledge is Dangerous

Little knowledge is dangerous, as they say. Even more dangerous is half baked language skills. Before I went as the Defence Attaché to Turkey, I was given a very basic exposure to Turkish language. Eight periods in eight weeks by a retired professor of Turkish history. My language skills were less than rudimentary.

Turks are very nationalist and proud people. They take pride in speaking Turkish even if they know English. I spoke to Turks in Turkish for better integration, never mind my skill level. They realized that and would accept my mistakes.

There also was a Defence Attaché from Turkish Cyprus. We were good friends and got along well. He spoke decent English and did not insist on speaking Turkish unless the interlocutor wanted to. I met him during a reception after a long time.

I chose to speak to him in Turkish.

"Cok zaman, size bakmadim, Ozkan Bey," (Long time did

not see you, Ozkan), I said.

"Evet, Ben tatille gitti." (Yes, I was on vacation).

"Cok Guzel,"(Damn good), I said.

"----------------" replied Ozkan in Turkish.

"Masha-allah," I said.

"----------------" replied Ozkan in Turkish again.

"Cok Guzel," (How wonderful), I said again.

Ozkan must have been wondering something is wrong with me. He said in English, "You are not understanding what I am saying." I said, "Really!" He then translated the conversation between us.

"You said that you had not seen me for a while. I said, I had gone on vacation. And you said, damn good. Then I said, I had gone because my sister had died and you said, excellent. Then I said, we buried her yesterday and you said, how wonderful."

Well, what was there for me to do? I tendered my sincere apologies, explaining I was only trying to speak his language. Fine man that he was, he said that was ok.

Little knowledge is indeed dangerous. I realized that I needed to work much harder on my Turkish.

Tailpiece

Well, that was the lighter side of life in the Indian Army. Interestingly, humour is recognized as one of the leadership traits in the army. That does not mean that every comedian from Bollywood would become a General. Nevertheless, there is no denying that a leader with a sense of humour keeps his command cheerful and in good spirits. Very often, the line of duty brings you face to face with a tough situation and the best way out of the tight corner is humour. I was posted in Turkey as the Defence Attaché. During a lunch with Turkish Naval officers, my hand hit the wine glass spilling red wine on the white table cloth, making a mess of it. Very embarrassing indeed. I thought quickly and told the three star Admiral, "As you can see sir, I leave my mark wherever I go." Everyone had a hearty laugh and the whole thing was forgotten as quickly as it had happened.

The subordinates feel comfortable with a zestful senior who appreciates the lighter side of life and differentiates between life before and after parade.

I was doing the senior command course in Mhow. The course finished and we were hosting a cocktail for the instructors. One instructor, a Colonel, walked up to us with a glass of 'nimbu paani' (lemonade). I offered him a drink. He said, "Thanks, I don't drink." I then said, "We are the hosts today, you won't pay for it." He said, "I still do not drink." I then said jokingly, "Sir, then you are wasting a vacancy in the army." He got a bit annoyed and said, "Yes, I am wasting a vacancy in the army," and walked away. Well, I may have crossed my brief but the seniors also need to be large hearted enough to accept a bit of leg pulling. That will go a long way in creating that wonderful environment.

Serving in a formation headquarter in Ambala as a Colonel, one youngster walked up to me in Sirhind club and introduced himself. As we chatted he asked me, "Sir, are you re-employed here." I had a hearty laugh and said, "No, I have another six years to go, but thanks for telling me how young I look and that I need to give up on competing with you in impressing young girls." The episode became famous in the cantonment. Later, when I was approved for Brigadier, the same youngster called me up and said, "Sir, heartiest congratulations. I believe you will soon be joining the gang of freedom fighters," (officers who live off the land). I thanked him, shared his joke and promised to be different.

As the Sector Commander in Lolab, I had an officer from Rajputana Rifles doing the Camp Commandant, a hardworking officer with a 'Can do' attitude. Having missed all his promotion boards, he was a perfect example of a man with the right attitude. He was grateful for all that the army and life had given him, despite facing a court martial. In Lolab, he saw life as a paid holiday. Jimmy in my battalion was a similar officer. Nothing could dent his enthusiasm and spirit.

While commanding my battalion, a subaltern asked me, "Sir, you are one of the few commanding officers who is smiling all the time." "Well," I said, "It is all about attitude. If your attitude is right, life is fun, anywhere - anytime. I for one choose to see the better side of the coin, every single time."

Well, when I missed my first board to Brigadier, I saw many opportunities there. One such opportunity was to leave my mark, yet once more (like the one on the lunch table in Turkey!). I resolved to be the best superseded officer the Indian Army had ever seen, and believe me, that was fun. Anyone that I spoke to was surprised by my concept. I guess, I commanded as much respect then, as I ever did, if not more. I downloaded "Auld lang Syne" as my signature tune in my mobile, for the old times sake, which were simply wonderful. Why am I writing all this? Because, unfortunately the way the system is, a very large part of the officer cadre will be superseded despite their competence. Majority of them, rather early in life. It is both necessary for the army and the officers themselves to develop the right attitude. Immensely disheartening, as supersession is, their and their family's happiness lies in developing that positive attitude and seeing things in a larger perspective of life that goes beyond ranks. On the balance sheet of life, we are more fortunate than most on this earth. There is a lot to live happily for, even after supersession - your pride - on which depends your happiness and the happiness of people who matter to you.

Change your attitude...
And you change your Life.

Adversity is a part of everyone's life and a good attitude helps you address your disappointments better.

Smile and laughter are God's special gift to mankind. They are infectious and spread happiness without taking away yours.

It is time to say goodbye. The journey of this book that started in 1992 has come to an end. I am sure it gave you an insight into a soldier's life and gave you a reason to smile.

God bless you all
Pahla Hamesha Pahla